PHYSICS PROBLEMS FOR
PROGRAMMABLE CALCULATORS

WAVE MOTION, OPTICS
AND MODERN PHYSICS

J. RICHARD CHRISTMAN
U.S. Coast Guard Academy

JOHN WILEY & SONS   NEW YORK
CHICHESTER BRISBANE TORONTO
SINGAPORE

PREFACE

This book is a continuation of PHYSICS FOR PROGRAMMABLE CALCULATORS: MECHANICS AND ELECTROMAGNETISM. It deals with wave motion, geometrical and wave optics, special relativity, and introductory quantum mechanics. There is also a chapter containing several topics in solid state and nuclear physics.

The material is meant to supplement the books by Halliday and Resnick: PHYSICS and FUNDAMENTALS OF PHYSICS. The chapters dealing with modern physics can also be used in conjunction with the two books by Resnick: INTRODUCTION TO SPECIAL RELATIVITY and RELATIVITY AND EARLY QUANTUM THEORY. All of these books are published by John Wiley & Sons.

Sections on special relativity and on integration of the Schrödinger equation might also form part of an intermediate modern physics course.

All of the problems involve straightforward applications of the basic principles covered in an introductory course. They represent opportunities for students to learn numerical techniques and apply them to the study of physical phenomena.

Many of the problems are designed to illuminate physical principles. The transport of energy and momentum by waves and the reflection of waves at boundaries are investigated. The probabilistic nature of phenomena at the quantum mechanical level is demonstrated for both photons and electrons. The probability that a particle reaches selected regions of space is calculated, then the place where the particle actually goes in a particular instance is found by using the capability of the calculator to select numbers randomly. The uncertainty principle is investigated by means of detailed calculations. Its experimental basis is emphasized by a calculation of the mean and standard deviation of coordinate data generated by the calculator.

The photoelectric effect, the Compton effect, and electron-positron annihilation are studied using the conservation laws for momentum and energy. Use of the calculator allows the study to include the role of the nearby heavy particle in the photoelectric effect and the influence of the initial electron momentum on the outcome of a Compton effect experiment.

For many physical phenomena, a graph which shows how some quantity varies is an important tool for understanding. In this book, much use is made of graphs. Wave functions, interference and diffraction patterns, probability distributions, electron densities in semiconductors, and the number of radioactive nuclei are all plotted as functions of relevant parameters. The calculator is used to take much of the drudgery out of graph plotting.

Some sections, particularly those dealing with standing waves, optics, and nuclear physics, can be used in connection with laboratory experiments. The values of various parameters given in the problems can be replaced by those actually encountered in the experiments.

In some cases, calculator projects can replace the more usual laboratory experiments. More time is then available for a deeper study of selected topics and this book provides a source of material. The chapters on special relativity and quantum mechanics are particularly adaptable to this type approach.

Programs are presented in the form of flow charts, which give the sequence of instructions to be followed by the machine. Instructions are written in a language similar to BASIC but, in most cases, they must be translated into the language understood by the machine. A brief review of the language used on the flow charts can be found in Appendix B. More details can be found in Chapter 1 of the volume on MECHANICS AND ELECTROMAGNETISM. The appendices of that volume contain information which is helpful in making the translation from the flow chart language to several commonly used languages.

Most problems are designed to produce answers with three significant figure accuracy. On occasion, when the result of a problem is a graph, accuracy is reduced to decrease running time. A few problems require answers which are accurate to more figures and note is made of these as they appear. It is tacitly assumed that numerical values given in the problems are accurate to the number of figures required even if a smaller number of figures is actually given.

I am particularly indebted to Robert Resnick and David Halliday for their helpful suggestions and strong encouragement. I am grateful to them for the opportunity to write these supplements to their textbooks.

I also wish to thank Saul Krasner, Gregory Cope, Bruce Russell, William Helgeson, Robert O'Hara, Michael Bray and Ellen Tossett for their help. Robert A. McConnin, Francine Fielding, and William Kellogg of Wiley provided much assistance in bringing this volume into being. Ruth Pflomm expertly typed the final manuscript. To all of these people go my grateful thanks.

I owe much to my family, who helped in a great many ways during the writing of this book. In appreciation, this book is dedicated to them: Mary Ellen, Stephen, and Karen.

New London, Connecticut 06320   J. Richard Christman
November 1981   U.S. Coast Guard Academy

TABLE OF PROGRAMS

TABLE OF CONTENTS

PHYSICS PROBLEMS FOR
PROGRAMMABLE CALCULATORS

Wave Motion, Optics
and Modern Physics

Chapter 17

WAVE PULSES

In this chapter, the programmable calculator is used to demonstrate some properties of waves. Although the principles can be applied to many different types of waves, the examples used here are all transverse waves on strings. Calculations concerning reflection at the string ends and the transfer of momentum and energy are discussed. This chapter supplements Chapter 19 of PHYSICS, Chapter 17 of FUNDAMENTALS OF PHYSICS, and material on wave motion found in other texts.

27.10.82

17.1   Pulses on a String

We consider waves which propagate along the x axis and take the function $y(x,t)$ to represent the form of the wave. For water waves, y represents the displacement of the water above (if positive) or below (if negative) the quiescent level. If the wave is electromagnetic, y might be any of the cartesian components of the electric or magnetic field. For sound waves, y gives the displacement of the moving particles, measured from their position in the absence of the wave. In each case, y is a function of both x and t. It is evaluated for some point with coordinate x and for some time t. y is a measure of the disturbance caused by the wave at the point with coordinate x, at time t.

It is possible to have waves which move in three dimensional space. A light wave, spreading out from some source, is an example. In that case, the disturbance function y is a function of three cartesian coordinates as well as a function of the time. Because of their mathematical simplicity and because they are easier to visualize, we concentrate on waves which move in one dimension only.

We consider waves on a stretched string. Originally the string is along the x axis and a disturbance is created by moving parts of it to the side, then letting go. The shape, which is the disturbance, moves along the string and $y(x,t)$ gives the displacement of the string at the point x and time t. y is measured from the original position of the string, the x axis.

No material particles travel with the wave. The string itself moves from one side to the other while the shape moves along the string. Each displaced portion of the string exerts a force on neighboring portions and these forces are just right to cause parts of the string in front of the wave to move and assume the correct shape. As the trailing edge of the wave passes, the string is pulled back toward its equilibrium position.

The speed of a wave on a string is determined by the tension F in the string and the linear mass density of the string. It is given by

$$v = (F/\mu)^{\frac{1}{2}} \tag{17-1}$$

where μ is the mass per unit length of the string.

If all parts of the wave move with the same speed, then the shape of the disturbance does not change as it moves along. This situation is easy to describe mathematically. Consider a string stretched along the x axis. At time t=0, the function f(x) gives the displacement of the string at the point with coordinate x. It tells how the string is originally displaced to start the wave. If the wave travels with speed v in the positive x direction, then

$$y(x,t) = f(x-vt) \tag{17-2}$$

gives the displacement of the string at the place x, for another time t.

This equation says that the displacement at the point x and time t is the same as the displacement at the point x-vt at time 0. The quantity vt, of course, is just the distance the wave travels in time t and the equation is a direct consequence of the supposition that the wave moves without change in shape.

If the functional form f(x) is given, y can easily be found. Everywhere x appears in f(x), replace it with the combination x-vt.

A program to evaluate y(x,t) for a given shape function f(x) is given and then the program is used to investigate some of the properties of waves.

The flow chart is shown in Fig. 17-1. The program itself is quite similar to the program of Fig. 2-4. For a given value of the time t, the program calculates f(x-vt) for a series of N values of x, starting with x_0 and separated by Δx. X1 contains x_0, X2 contains Δx, X3 contains N, and X4 contains the wave speed v. The function f(x), which gives the shape of the string at t=0, must be known.

303

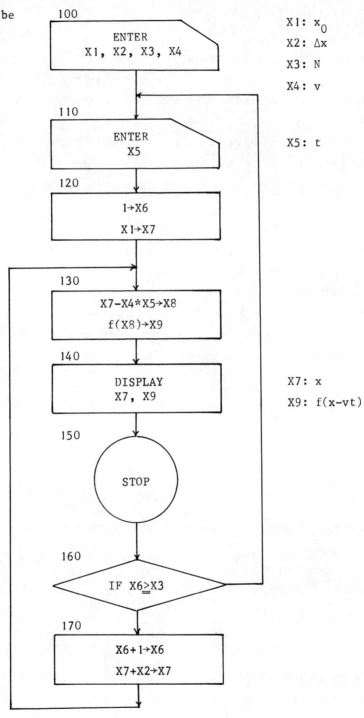

Figure 17-1.   Flow chart for a program to evaluate functions which describe traveling wave forms.

X1: x_0
X2: Δx
X3: N
X4: v

X5: t

X7: x
X9: $f(x-vt)$

At line 110, the time is entered and stored in X5. X6 is a counter used to count the intervals in x; at line 120 it is initialized to 1. X7 contains x; at line 120 it is initialized to x_0.

The loop over x is entered at line 130, where x–vt is calculated and stored in X8. The shape function is then evaluated and its value is stored in X9. Note that the argument of the function is assigned the value x–vt, the number stored in X8.

At line 140, x and f(x–vt) are displayed and the machine stops to allow these values to be copied. If the machine has printing capability, it is worthwhile to have it print the results rather than display them. The STOP instruction can then be omitted.

After restarting, the machine checks to see if N intervals have been considered. If they have, it returns to line 110 to accept another value of the time. If they have not, both x and the counter are incremented, x by Δx and the counter by 1. The machine then goes to line 130 to begin the evaluation of f(x–vt) for the new value of x.

The program is first used to show that y(x,t)=f(x–vt) does represent a shape which moves with speed v toward more positive x and that the shape does not change as it moves.

Problem 1.   Consider a string which initially has a shape given by

$$f(x) = 0.02 \, e^{-x^2/9}$$

where f and x are in meters. Suppose this pulse moves in the positive x direction with speed v=25 m/s.

a.   Use the program of Fig. 17–1 to plot y(x,t) for t=0, 0.2 s, and 0.4 s. Evaluation of the function at line 130 should proceed according to the instruction

.02*EXP(–X8↑2/9)→X9

Start with x=–5 m and plot points every 1 m until x=20 m is reached. If it becomes tedious to write down all the results, you may ignore those for which X9 is less than 0.001. Call them 0 for purposes of plotting.

b.   Notice that the pulse does move toward more positive x and that it does not change shape. The maximum of the pulse is easily identifiable. Check that the speed is

25 m/s by dividing the distance moved by the peak in the first 0.4 s by the time interval. Read the distance from the graphs.

A pulse with the same shape may propagate in the negative x direction. If f(x) gives the initial shape of the string, then for such a pulse,

$$y(x,t) = f(x+vt) \qquad (17\text{-}3)$$

gives the shape at time t. Here v is the speed of the wave and is positive. This equation is like Eqn. 17-2 except that the sign preceding v is changed. A + sign is used for a negative going pulse and a − sign is used for a positive going pulse.

Problem 2. Consider a string which initially has a shape given by

$$f(x) = 0.02\,e^{-x^2/9}$$

where f and x are in meters. Suppose the pulse moves in the negative x direction with speed v=25 m/s.

a. Use the program of Fig. 17-1 to plot y(x,t) for t=0, 0.2 s, and 0.4 s. Evaluation of the function at line 130 should proceed according to the instructions

X7+X4*X5→X8

.02*EXP(-X8↑2/9)→X9

Start with x=−20 m and plot points every 1 m until x=+5 m is reached. Again you may wish to copy only those values of y which are greater than 0.001 m.

b. Verify that the graph depicts a pulse which

i.   travels in the negative x direction,

ii. has a speed of 25 m/s,

iii. moves without change in shape.

Waves on a string, and others, obey a superposition principle. Two or more waves can exist simultaneously at the same place and, if they do, the displacement of the string is the sum of the displacements produced by the individual waves. In regions

306

where there is more than one wave, the net displacement can be larger than the displacement due to any of the component waves or it can be smaller, depending on the signs of the various contributions.

Problem 3. Consider a string which carries two pulses. The first initially has the form

$$f_1(x) = 0.02 \, e^{-x^2/9}$$

and travels in the positive x direction with speed v=25 m/s. The second initially has the form

$$f_2(x) = 0.02 \, e^{-(x-15)^2/9}$$

and travels in the negative x direction with the same speed. In these expressions, f_1, f_2, and x are in meters. Plot the displacement of the string for every 1 m from x=-5 m to x=+20 m. Make separate plots for each of the times t=0, 0.2 s, 0.3 s, 0.4 s, and 0.6 s. To do this, replace line 130 of Fig. 17-1 by

.02*EXP(-(X7-X4*X5)↑2/9)+.02*EXP(-(X7-15+X4*X5)↑2/9)→X9

X8 is not used.

     The two pulses move toward each other. When they meet, the displacement in the region of overlap becomes large. The pulses continue moving in the same directions and move away from each other. The shapes do not change.

Problem 4. Rework problem 3 but take the initial form of the second pulse to be

$$f_2(x) = -0.02 \, e^{-(x-15)^2/9}$$

where f_2 and x are in meters. The two functions have different signs, indicating that the displacements are in opposite directions. Plot the displacement of the string for every 1 m from x=-5 m to x=+20 m. Make separate plots for each of the times t=0, 0.2 s, 0.3 s, 0.4 s, and 0.6 s. To do this, replace line 130 of Fig. 17-1 by

$$.02*EXP(-(X7-X4*X5)\uparrow 2/9)-.02*EXP(-(X7-15+X4*X5)\uparrow 2/9)\rightarrow X9$$

The two pulses approach each other and, when they meet, the action of one tends to nullify the action of the other. For one value of the time, the displacement of the string is zero everywhere. Later, the pulses emerge from the interaction and continue on their ways, unchanged.

Waves are usually generated by moving one end of the string. We let the function g(t) represent the displacement of one end, presumed to be at x=0. This function tells how the end is moved in order to generate the wave. At time t, the displacement at some point x is the same as the displacement at the end but at the earlier time t-x/v. The time interval is just the time it takes the disturbance, traveling at speed v, to reach x. Thus

$$y(x,t) = g(t-x/v). \qquad (17\text{-}4)$$

We consider examples for which the end of the string is not set in motion until t=0 and the motion of the end stops at $t=t_f$. That is, g(t)=0 for t<0 and g(t) is a given constant A for $t>t_f$. The constant, of course, is given by $A=g(t_f)$. Line 130 should be replaced by the series of instructions

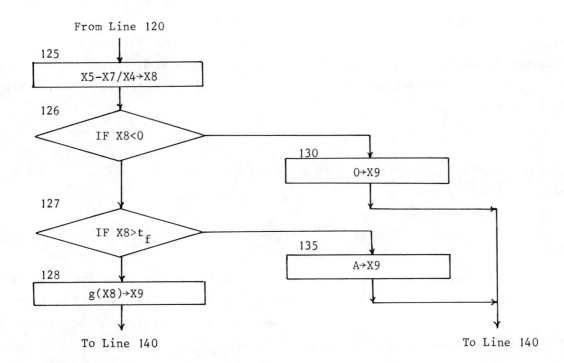

308

First t-x/v is computed and stored in X8. If this number is negative, then y, stored in X9, is set equal to 0. If the value in X8 is greater than t_f, y is set equal to A. For $0<t-x/v<t_f$, g(t-x/v) is calculated and stored in X9. Values for t_f and A must be supplied when the machine is programmed.

Problem 5.   Starting at t=0 and continuing for 0.2 s, the free end of a string, at x=0, is pulled up with a constant speed of 0.15 m/s. For $0 \leq t \leq 0.2$ s, g(t)=0.15t and, for t>0.2 s, g(t)=0.03 m. For each of the following two wave speeds, plot the displacement of the string as a function of position along the string. Make separate plots for each of the times t=0, 0.1 s, 0.2 s, and 0.3 s. Although the given conditions produce displacements which are not small, assume the wave travels at constant speed without change in shape.

a.   v=8 m/s.

b.   v=22 m/s.

The plots you have made show a taut string being moved from one position to another, parallel position. The shift in position is brought about by moving one end. The string does not move all at once, but rather the one end moves, and, in moving, creates a disturbance which travels along the string. The disturbance is a straight line distortion of the string. In front of the distortion, the displacement of the string is zero and, as the disturbance passes any given point, it leaves behind a string which is displaced 0.03 m from its original position. The speed with which the disturbance moves is the wave speed for the string. It depends on the tension in the string and the linear mass density of the string and is unrelated to the speed with which the free end is pulled.

Problem 6.   Starting at t=0 and continuing for 0.6 s, the free end of a string at x=0, has a displacement given by

$$g(t) = 0.03 \sin(\frac{\pi}{0.3} t).$$

Here g is in meters, t in seconds, and the argument of the sine function is in radians. Before and after the motion, the end is not displaced. Take the wave speed to be 10 m/s and, for each of the times t=0, 0.4 s, 0.8 s, and 1.2 s, plot the displacement of the string as a function of position along the string.

At any given point on the string, the displacement is that which was created at an earlier time at the end. When the string end stops moving, the disturbance is detached from the source at the origin and continues to move along the string.

Once detachment occurs, the disturbance moves without change in shape and with constant speed along the string, just like the pulses of previous problems. A displaced portion of the string exerts a force on neighboring portions and this force causes the neighboring portions to distort in just the right way to reproduce the wave shape a little further along the string.

The plots you made for the preceding two problems are somewhat unrealistic since the string end has infinite acceleration when it starts and stops. The resulting pulse has sharp corners. Nevertheless, these plots show the chief features of wave creation.

The pulses we have been considering continue to travel to the other end of the string. What happens there is examined in the next section.

17.2   Reflections at End Points

When a pulse gets to the end of the string, a pulse traveling in the opposite direction is generated by the motion of the string at the end. The original pulse is said to be reflected.

Two special cases are of particular interest: the end of the string is fixed and does not move, and the end of the string travels from side to side with no transverse force acting on it except the force of the neighboring piece of string. The latter situation arises if a loop is made in the string and the loop slides along a frictionless dowel, perpendicular to the unstretched string.

First consider a string with the end at x=L fixed and let $y_1(x,t)$ be a pulse incident from the left. Let $y_2(x,t)$ be the reflected pulse. The displacement of the string is given by $y=y_1+y_2$ and this must vanish at x=L for all values of the time. In mathematical terms,

$$y_1(L,t) + y_2(L,t) = 0 \qquad\qquad (17\text{-}5)$$

310

so

$$y_2(L,t) = -y_1(L,t).$$ (17-6)

This expression gives y_2 at the fixed end. Neither y_1 nor y_2 vanish at the end but their sum does.

We wish to find y_2 for other places on the string. The role of the fixed end point is to produce a motion which exactly cancels $y_1(L,t)$ and y_2 is exactly the same as the pulse produced when the end of the string is displaced according to $-y_1(L,t)$, a function of time. A general point x is L-x from the end and it takes the reflected pulse the time (L-x)/v to travel from the end to x. At time t the displacement at x, due to the reflected wave, is the same as the displacement at the end but at time t-(L-x)/v. So

$$y_2(x,t) = -y_1(L,t-(L-x)/v).$$ (17-7)

To obtain this result, t is replaced by t-(L-x)/v in $-y_1(L,t)$. If f(x) is the initial shape function for the incident pulse, then $y_1(x,t)=f(x-vt)$ and $y_2(x,t)=$ $-f(L-vt+L-x)=-f(2L-x-vt)$. Here x is replaced by L and t is replaced by t-(L-x)/v. Notice that y_2 is a function of x+vt and is therefore a pulse which travels in the negative x direction. The total displacement is

$$y(x,t) = f(x-vt) - f(2L-x-vt).$$ (17-8)

If the initial shape function is given as a function of x, the first term in the above expression is found by substituting x-vt for x and the second term is found by substituting 2L-x-vt for x.

We study the reflection of a pulse from the fixed end of a string by plotting the displacement as a function of position for various times. These plots give a sequence of pictures of the string as the pulse approaches the end and is reflected.

We use the same pulse as we used in the first four problems of the last section so that you can compare the displacement of the string when the end is fixed to the displacement, plotted previously, when reflection is not considered.

<u>Problem 1.</u>   A pulse, with initial shape given by

$$f(x) = 0.02 \ e^{-(x-15)^2/9},$$

is incident from the left on the end of a string fixed at x=20 m. In this expression, f and x are in meters. Take the wave speed to be 13 m/s and plot y(x,t) as a function of x for each of the times t=0, 0.1 s, 0.2 s, 0.3 s, 0.4 s, 0.5 s, and 0.6 s. The function to be evaluated at line 130 is

$$0.02 \ e^{-(x-vt-15)^2/9} - 0.02 \ e^{-(2L-x-vt-15)^2/9}.$$

The influence of the fixed end is evident on the graphs. The pulse becomes blunt nosed as it approaches the end. This occurs, of course, because the front edge has been reflected and there is now a left going, negative pulse which, when added to the incident pulse, reduces the displacement near the end point. By t=0.4 s, slightly more than half the incident pulse has been reflected and, at any point on the string, the reflected pulse has a slightly greater magnitude than the incident pulse. The two pulses almost cancel each other but there is a small, negative, residual displacement.

We now consider the situation when the end of the string is not fixed but is allowed to move freely from side to side. There is again a reflected pulse but the displacement associated with it is in the same direction as the displacement associated with the incident pulse.

Because there is no string beyond x=L to exert a restoring force on the string at x=L and because the end of the string is light, the displacement at the end is greater than that produced by the incident pulse alone. Again the source of the reflected wave is the motion of the end associated with the incident pulse, considered as a function of time. In mathematical terms, if $y_1(x,t)$ is the incident pulse, then

$$y_2(x,t) = +y_1(L,t-(L-x)/v) \qquad\qquad (17\text{-}9)$$

is the reflected pulse. If f(x-vt) is the shape function for the incident pulse,

312

then f(2L-x-vt) is the shape function for the reflected pulse. This form is
obtained by substituting L for x and t-(L-x)/v for t in the functional form for the
incident pulse. The total displacement of the string is given by

$$y(x,t) = f(x-vt) + f(2L-x-vt).$$ (17-10)

At x=L, the string remains parallel to the x axis as it moves. It should
be clear to you that Eqn. 17-10 obeys $\partial y/\partial x=0$ for x=L no matter what the functional
form of f and no matter what the value of t. Since the string is parallel to the
x axis at the end, the transverse component of the force exerted on the end by
neighboring portions of the string is zero. The mass at the end is vanishingly
small, so Newton's second law is obeyed as long as the transverse component of the
string's acceleration remains finite. If the string were not parallel to the x axis,
there would be a transverse force acting on a vanishingly small mass and the result
would be an infinite acceleration.

Again we study the reflection process by plotting the displacement of the
string for a sequence of times as the incident pulse approaches the end and is
reflected.

Problem 2.   A pulse initially has the form

$$f(x) = 0.02\, e^{-(x-15)^2/9}$$

with f and x in meters. It is incident from the left on the free end of a string
at x=20 m. The speed of the pulse is 13 m/s. Plot y(x,t) as a function of x for
every 0.2 s from t=0 to t=0.8 s.

The flapping of the end is obvious in these plots. At t=0, the pulse has
just reached the end and the end has started to wave. At t=0.2 s, the end is almost
at the same displacement as the peak of the pulse although the peak has not yet
reached the end point. By t=0.4 s, the peak has been reflected and the end has
moved to a point which is almost double the peak displacement of the incident pulse.
In fact, at this time the end has already reached its maximum displacement and is

starting back toward the x axis again. The plots for t=0.6 s and t=0.8 s show the end returning to zero displacement. The reflected pulse has now been formed and is moving off to the left.

17.3   Transport of Energy and Momentum

As a pulse moves along the string, energy and momentum are carried from one place to another. This transport of energy and momentum is obvious. Look at the graphs plotted in connection with problem 1 of section 17.1. At t=0, the string in the neighborhood of x=0 is moving and that portion of the string has a certain amount of kinetic energy. Later, that part of the string is motionless and a part of the string at larger x is moving. The kinetic energy has moved, with the pulse, from smaller to larger x.

There is also a potential energy associated with the pulse. As the string is displaced from its equilibrium position, it must elongate and the potential energy is associated with the work done by the tension during the elongation process. With the motion of the pulse, different parts of the string are deformed at different times and the potential energy moves with the pulse.

The same statements can be made about the momentum. As the pulse moves into a segment of the string, that segment acquires momentum. Later, as the pulse moves out, the momentum is lost to a neighboring region. The momentum of a small segment of the string is, of course, the mass of that segment multiplied by its velocity and the momentum is in the direction of motion of the string segment. For the pulses considered here, it is perpendicular to the direction of motion of the pulse.

The calculator can be used to help demonstrate the transport of energy. We consider a small segment of the string, with length Δx, and calculate the total energy of that segment. We then add the contributions of all segments of the string to find the total energy in the string. Since we neglect frictional forces exerted by one part of the string on other parts, we expect the total energy to remain constant as time goes on. The energy moves, undiminished, from one place to another.

If μ is the mass per unit length of the string, the segment of length Δx at x contains mass $\mu \Delta x$. The string there moves with velocity $\partial y / \partial t$, so its kinetic energy is $K = \frac{1}{2} \mu \Delta x (\partial y / \partial t)^2$. Here the derivative is evaluated at the center of the segment. If Δx is sufficiently small, all parts of the segment move with nearly the same velocity. The total kinetic energy is found by summing the contributions of all segments or, in the limit of infinitesimal segments, by evaluating the integral

$$K = \frac{1}{2} \mu \int_0^L (\partial y / \partial t)^2 \, dx. \qquad (17\text{--}11)$$

When the string is displaced, it elongates. If Δy is the change in the displacement from one end to the other of a string segment, the length of the segment is $\left[(\Delta x)^2 + (\Delta y)^2 \right]^{\frac{1}{2}}$ or, if Δx is factored out, $\left[1 + (\Delta y / \Delta x)^2 \right]^{\frac{1}{2}} (\Delta x)$ and the elongation is the difference between this and the length Δx of the undisturbed segment. That is, the elongation is given by $\left[1 + (\Delta y / \Delta x)^2 \right]^{\frac{1}{2}} (\Delta x) - (\Delta x)$. If $\Delta y / \Delta x$ is small, the binomial theorem in the form $(1+A)^{\frac{1}{2}} = 1 + \frac{1}{2} A + \ldots$ can be used to approximate the expression for the elongation. The elongation is then given by $\frac{1}{2} (\Delta y / \Delta x)^2$. This elongation is brought about by the tension F in the string and, since this is uniform and tangent to the string, the work it does is given by $\frac{1}{2} F (\Delta y / \Delta x)^2 \Delta x$. This work changes the potential energy of deformation by a like amount.

For a sufficiently short segment, the ratio $\Delta y / \Delta x$ is taken to be the partial derivative $\partial y / \partial x$, evaluated at the position of the segment. Thus, for the segment $dU = \frac{1}{2} F (\partial y / \partial x)^2 \, dx$ and, for the whole string,

$$U = \frac{1}{2} F \int_0^L (\partial y / \partial x)^2 \, dx. \qquad (17\text{--}12)$$

The program of Fig. 8-1 (Appendix C) can be used to evaluate the integrals in Eqns. 17-11 and 17-12. The function which is evaluated at lines 130, 170, and 180 is either $(\partial y / \partial t)^2$ or $(\partial y / \partial x)^2$, depending on the quantity to be calculated.

Problem 1. Suppose the pulse of problem 1, section 17.1, is moving on a string with linear mass density $\mu = 0.08$ kg/m.

a. What is the tension in the string? (Ans: 50 N)

b. Show that the kinetic energy in the pulse at time t is given by

$$K = \tfrac{1}{2}(0.04/9)^2 \, \mu v^2 \int (x-vt)^2 \, e^{-(x-vt)^2/4.5} \, dx.$$

c. Show that the potential energy in the pulse at time t is given by

$$U = \tfrac{1}{2}(0.04/9)^2 \, F \int (x-vt)^2 \, e^{-(x-vt)^2/4.5} \, dx.$$

Since $F = \mu v^2$, the kinetic and potential energies are numerically the same.
This result is generally true for waves which have the form f(x-vt) or f(x+vt).

d. Use the program of Fig. 8-1, with N=50, to evaluate the integrals for t=0 and
t=1 s. For t=0, integrate from x=-10 m to x=+10 m and, for t=1 s, integrate
from x=+15 m to x=+35 m. These integration limits are chosen so that, in
each case, the integral includes all except the very small tails of the pulse.
Evaluate the kinetic energy, the potential energy, and the total mechanical
energy for t=0 and t=1 s. Note that each of the energies has the same value
at the two times. The energy has simply moved from the region around x=0 to
the region around x=25 m.

It is possible to calculate the energy passing a given point on the string.
We consider a pulse moving to the right along the x axis, pick a point x on the
string, and calculate the energy which passes that point per unit time interval.
Energy is transferred from one segment to a neighboring segment by the action of
the force exerted by the first segment on the second. When the string just to the
right of x moves through an infinitesimal displacement dy, the energy it receives
is F_y dy, where F_y is the transverse component of the force exerted by the string
just to the left of x.

Now $F_y = -F\sin\theta$ where F is the tension in the string and θ is the angle
between the string and the x axis. For small displacements $\sin\theta \approx \partial y/\partial x$ and the
energy received is $dE = -F(\partial y/\partial x) \, dy$. The energy passing the point x per unit time
is

$$P = dE/dt = -F \, (\partial y/\partial x)(\partial y/\partial t), \qquad\qquad (17\text{-}13)$$

where the symbol P is used to denote the power transmitted.

In the time interval from t_0 to t_f, the total amount of energy which passes the point at x is given by

$$E = \int_{t_0}^{t_f} P \, dt = -F \int_{t_0}^{t_f} (\partial y/\partial x)(\partial y/\partial t) \, dt. \qquad (17\text{-}14)$$

Problem 2.   In the time interval from t=0 to t=1 s, the pulse of problem 1, section 17.1, moves from the neighborhood of x=0 to the neighborhood of x=25 m. Essentially all of the energy in it passes the point x=12.5 m in that time interval. Use Eqn. 17-14 to calculate this energy.

a.   Show that, for the pulse considered, the total energy which has passed the point at x is

$$E = (0.04/9)^2 \, Fv \int_0^1 (x-vt)^2 \, e^{-(x-vt)^2/4.5} \, dt.$$

b.   Put x=12.5 m and use the program of Fig. 8-1, with N=50, to evaluate the integral. Use the value of F you found in problem 1 of this section to calculate the energy E which passes x=12.5 m during the time interval from t=0 to t=1 s. Compare the answer to the total energy in the pulse.

Notice that Eqn. 17-13 gives the total mechanical energy transferred across the selected point per unit time. Part of this energy is in the form of kinetic energy and part is in the form of potential energy. The work done by the segment on the left changes both the kinetic and potential energies of the segment on the right.

Look at the result for part a of problem 2 above and note that P=(K+U)v. This is a general relationship for waves and it indicates that the energy moves with the wave speed v for waves which do not change shape.

As demonstrated in problem 4 of section 17.1, it is possible to have two pulses traveling in opposite directions on the same string so that, at some instant of time, the displacement everywhere is zero. At that instant, the string is not deformed and the potential energy vanishes. The string is still moving, however, and the energy is entirely in the form of kinetic energy.

Problem 3. Consider the string and pulses of problem 4, section 17.1. Take the linear mass density to be 0.08 kg/m. Use the program of Fig. 8-1, with N=50.

a. Calculate the kinetic, potential, and total mechanical energies in each pulse at t=0, when they are far apart. You will need to compute the tension in the string using $F=\mu v^2$. For each pulse, take the integration limits to be 10 m on either side of the peak.

b. Calculate the kinetic energy of the string at time t=0.3 s, when y=0 everywhere. To do this, you must take y to be the total displacement, the sum of the displacements of the individual pulses. The derivative of this function is squared to form the integrand in the expression for the kinetic energy.

The same situation occurs when a symmetric pulse is reflected from the fixed end of a string. For the pulse of problem 1, section 17.2, y=0 everywhere at t=0.385 s. The energy has not disappeared but is entirely kinetic in nature.

It is clear for the combined pulses that the kinetic and potential energies are not numerically equal, as they were for the pulse of problem 1 in this section. This does not violate the statement made following that problem since the displacement of the string cannot be described by f(x-vt) or f(x+vt).

The same analysis can be applied to the momentum. A string segment of length Δx has momentum $p=\mu(\partial y/\partial t) \Delta x$ and the total momentum of a finite portion of the string is given by

$$p = \mu \int (\partial y/\partial t)\ dx. \qquad (17\text{-}15)$$

If this integral is evaluated over an entire pulse, the result is zero since the string in front and behind the pulse is not moving.

318

Momentum is transferred from one segment to another by virtue of the force of the first on the second. For a pulse going left to right, the momentum which passes a selected point per unit time is equal to the transverse component of the force exerted by the string to the left of the point on the string to the right of the point. That is,

$$dp/dt = -F(\partial y/\partial x) \qquad (17-16)$$

and the total momentum which passes the point at x during the time interval from t_0 to t_f is given by

$$p = -F \int_{t_0}^{t_f} (\partial y/\partial x) \, dt. \qquad (17-17)$$

Problem 4. Consider the pulse of problem 1, section 17.1. Take $\mu=0.08$ kg/m. The total momentum in this pulse is zero so it is not worthwhile to study the flow of total momentum. We can, however, watch what happens to the momentum in part of the pulse, say the right half.

a. For t=0, use the program of Fig. 8-1, with N=50, to calculate the total momentum in that portion of the string which lies between x=0 and x=10 m. At that time, the peak is at x=0.

b. For t=1 s, use the program of Fig. 8-1, with N=50, to calculate the total momentum in that portion of the string which lies between x=25 m and x=35 m. The peak is now at x=25 m.

c. Calculate the momentum which passes the point x=12.5 m in the time interval from t=0 to t=0.5 s. This is the time it takes for the peak to move from x=0 to x=12.5 m. Use Eqn. 17-17 and the program of Fig. 8-1, with N=50.

The energy and momentum in the pulse must be supplied by the mechanism which creates the pulse. For example, if the pulse is created by moving one end of the string, work is done by the external force causing the motion. Since the external force acts on a vanishingly small segment of the string with vanishingly small mass, the external force has the same magnitude as the tension in the string and it acts along the line tangent to the string at the end. The total energy supplied by the source is given by Eqn. 17-14, evaluated for the end of the string and the total momentum supplied by the source is given by Eqn. 17-17, also evaluated for the end of the string.

These expressions can be written in terms of the function g(t), which describes the displacement of the string end as a function of time. In time dt the displacement at the end changes by (dg/dt) dt. At the end of the time interval, the point at x=v dt has the same displacement as the point at the end had at the beginning of the interval. So the slope of the string at its end is given by $(\partial y/\partial x)$ = -(dg/dt) dt/(v dt)=-(dg/dt)/v. The external force supplies energy at the rate

$$P = (F/v)(dg/dt)^2 \qquad\qquad (17-18)$$

and momentum at the rate

$$dp/dt = (F/v)(dg/dt). \qquad\qquad (17-19)$$

These expressions result when $\partial y/\partial t$=dg/dt and $\partial y/\partial x$=-(dg/dt)/v are substituted into Eqns. 17-13 and 17-16.

Problem 5. Consider the situation described in problem 6, section 17.1. Take μ=0.08 kg/m.

a. Calculate the total energy supplied by the source during the first 0.4 s. Use Eqn. 17-18 and the program of Fig. 8-1, with N=50.

b. Calculate the total energy in the pulse at t=0.4 s. To do this, evaluate Eqns. 17-11 and 17-12 using integration limits of x=0 and x=4. This is the extent of the pulse at t=0.4 s. Compare your answer with that of part a.

c. Calculate the total momentum supplied by the source during the first 0.4 s. Use the program of Fig. 8-1, with N=50. The integrand, of course, is the right side of Eqn. 17-19.

d. Calculate the total momentum in the pulse at t=0.4 s. Use Eqn. 17-15 with integration limits of x=0 and x=4 m. Compare your answer with that of part c.

Chapter 18
SINUSOIDAL WAVES

The functional form of traveling sinusoidal waves is studied, with particular emphasis on the concepts of amplitude, frequency, wavelength, and wave number. Beat phenomena are explored in several problems and a section on standing waves is included. Transverse waves on strings are used as examples. The material is designed to augment Chapters 19 and 20 of PHYSICS, Chapters 17 and 18 of FUNDAMENTALS OF PHYSICS, and similar sections of other texts.

18.1   Sinusoidal Waves

If one of the ends of a taut string is caused to move in simple harmonic motion, then a sinusoidal wave is created on the string. The displacement has one of the two forms

$$y(x,t) = y_m \sin(kx - \omega t + \phi) \qquad \text{(right going)} \qquad (18\text{-}1)$$

or

$$y(x,t) = y_m \sin(kx + \omega t + \phi), \qquad \text{(left going)} \qquad (18\text{-}2)$$

where y_m, k, ω, and ϕ are constants. Here the positive x direction is toward the right.

At any instant of time the string is in the shape of a sine function. The maximum displacement is y_m and this quantity is called the amplitude of the wave. It is a positive number.

If we seek points where the string has its maximum displacement for any time t, we would find a series of such points, spaced at uniform intervals along the string. The interval width is called the wavelength and it is related to the wave number k, which multiplies x in Eqn. 18-1 and Eqn. 18-2. It is denoted by λ.

For two coordinate points which differ by a wavelength, the argument of the sine function is different by 2π radians. That is, if $x_2 - x_1 = \lambda$ then $k(x_2 - x_1) = 2\pi$

so

$$\lambda = 2\pi/k. \qquad (18\text{-}3)$$

The wave number k has the unit of reciprocal distance and the wavelength has the unit of distance.

Every point on the string oscillates in simple harmonic motion. The quantity ω, which multiplies t in Eqns. 18-1 and 18-2, is the angular frequency and is measured in radians/s. For a fixed value of x, the displacement at the end of a time interval $2\pi/\omega$ is the same as at the beginning. This repeat time is called the period of the oscillation and is denoted by T:

$$T = 2\pi/\omega, \qquad (18\text{-}4)$$

a relationship which follows from Eqn. 18-1 or 18-2. For a fixed value of x, if times t_1 and t_2 differ by T, then ωt_1 and ωt_2 differ by 2π and Eqn. 18-4 follows.

Another quantity associated with the oscillation of the string is the frequency ν. It gives the number of oscillations which occur at any point in a unit time interval (1 s, for example), and it is the reciprocal of the period:

$$\nu = 1/T = \omega/2\pi. \qquad (18\text{-}5)$$

The frequency has the unit of reciprocal time and 1 cycle per second is called 1 hertz (Hz).

The argument of the trigonometric function is called the phase. For the wave of Eqn. 18-1, it is given by $kx-\omega t+\phi$. The angle ϕ is called the phase constant and it can be found, for example, by examining the displacement at x=0 and t=0:

$$y(0,0) = y_m\sin(\phi). \qquad (18\text{-}6)$$

For a given y(0,0) and y_m, there are two angles which satisfy this relationship and the correct one can be chosen if the string velocity at x=0, t=0 is known:

$$\left(\partial y/\partial t\right)_{\substack{x=0 \\ t=0}} = -y_m\omega\,\cos(\phi). \tag{18-7}$$

In practice, all that need be known is the sign of the string velocity.

Both of the wave forms, Eqns. 18-1 and 18-2, move with a wave speed given by

$$v = \omega/k = \lambda/T. \tag{18-8}$$

Since $kx-\omega t=k(x-vt)$ and $kx+\omega t=k(x+vt)$, the functions given by these equations have the forms $f(x-vt)$ and $f(x+vt)$ respectively, and they represent traveling waves. The waves move without change in shape and they carry energy and momentum from one part of the string to another.

Problem 1.   Use the program of Fig. 17-1 to plot the sinusoidal wave

$$y(x,t) = 0.2 \sin 20(x-350t)$$

as a function of x

a.   Plot points every 0.02 m from x=0 to x=0.6 m. Make separate graphs for t=0 and for $t=2\times10^{-4}$ s.

b.   Consider the graph for t=0 and measure the distance from one maximum to a neighboring maximum. Compare the result with the value for the wavelength computed using k from the above equation.

c.   Using the two graphs, measure the distance moved along the string by one of peaks in 2×10^{-4} s. Assume the wave moves in the positive x direction and that the smallest possible displacement is the correct one. Use this distance to calculate the speed of the wave and compare your answer with ω/k. Take values of ω and k from the functional form of the wave given above.

d.   Suppose all you know about the wave is the information given by the two graphs. Is there any way you can tell which way the wave is moving or what the speed is? Write down equations giving y(x,t) for the following waves, all of which reproduce your graphs for t=0 and $t=2\times10^{-4}$ s.

i. A wave going in the positive x direction with a greater speed than the given wave.

ii. A wave going in the negative x direction.

For other values of the time, of course, the waves are different.

Problem 2. Consider a right going sinusoidal wave of the form given by Eqn. 18-1.

a. Show that the power transmitted across the point at x is given by

$$P = Fk\omega y_m^2 \cos^2(kx-\omega t+\phi).$$

b. Consider the wave of problem 1 and take the linear mass density of the string to be 0.05 kg/m. Calculate the energy transmitted across the point x=7 m during ¼ cycle, ½ cycle, and 1 cycle, with each time interval starting at t=0.02 s. You will need to calculate the tension in the string, using $v=\sqrt{F/\mu}$.

c. For the wave of part b, calculate the momentum transmitted across the point x=7 m during ¼ cycle, ½ cycle, and 1 cycle, with each time interval starting at t=0.02 s.

d. The average power transmitted is defined to be the energy transmitted during 1 cycle, divided by the period. What is the average power transmitted by the wave of part b?

18.2 Beats

Two waves, moving in the same direction and having nearly the same frequency, are said to produce a beat. We add the two waves

$$y_1(x,t) = y_m\sin(k_1x-\omega_2t) \tag{18-9}$$

and

$$y_2(x,t) = y_m\sin(k_2x-\omega_2t). \tag{18-10}$$

Since both waves are on the same string, they have the same speed and $\omega_1/k_1=\omega_2/k_2=v$.

The trigonometric identity sin(A+B)=sinA cosB + cosA sinB and some algebraic manipulation can be used to show that

$$y_1 + y_2 = 2y_m \cos\left[\frac{k_1-k_2}{2}x - \frac{\omega_2-\omega_1}{2}t\right] \sin\left[\frac{k_1+k_2}{2}x - \frac{\omega_1+\omega_2}{2}t\right]. \qquad (18\text{--}11)$$

This sum can be interpreted as a sinusoidal wave with angular frequency and wave number equal to the appropriate averages of the values for the two component waves: $k=(k_1+k_2)/2$ and $\omega=(\omega_1+\omega_2)/2$. The amplitude then varies with x and t.

Eqn. 18-11 is true no matter what the values of ω_1 and ω_2. The beat phenomenon, however, occurs only when $\omega_2-\omega_1$ is much smaller in magnitude than both ω_1 and ω_2. Then the average angular frequency is nearly the same as the frequency of either component and the amplitude is a slowly varying function of time. $(\omega_2-\omega_1)$ is called the angular beat frequency.

If the waves are acoustic and in the audible frequency range, then the pitch that is heard corresponds to the average frequency. The sound grows louder, then softer as the long period amplitude factor grows and diminishes. It is as if a radio produces a tone with angular frequency $(\omega_1+\omega_2)/2$ and the volume control is turned first one way, then the other, with angular frequency $(\omega_2-\omega_1)$.

Problem 1.   Use the program of Fig. 17-1 to plot

$$y(x,t) = 0.2\sin(20x-3000t) + 0.2\sin(22x-3300t)$$

as a function of t for x=0. The two sinusoidal waves have nearly the same frequency and the composite wave exhibits a beat. Here y and x are in meters, t is in seconds.

a.   Plot points every 5×10^{-4} s from t=0 to t=0.04 s.

b.   The rapidly varying component of the graph should have a frequency equal to the average of those of the two component waves. On the graph, measure the time T_1 between successive peaks and calculate $\omega=2\pi/T_1$. Compare the result with $(\omega_1+\omega_2)/2=3150$ radians/s.

c.   On the graph, measure the period T_2 of the envelope and calculate $2\pi/T_2$. Compare the result with $(\omega_2-\omega_1)=300$ radians/s. It should be clear why this, not $(\omega_2-\omega_1)/2$, is the beat frequency.

If ω_1 and ω_2 are not nearly the same, Eqn. 18-11 still holds, but now the interpretation as a beat phenomenon is harder to make.

Problem 2. Take y and x to be in meters, t to be in seconds, and plot

$$y(x,t) = 0.02\sin(20x-3000t) + 0.02\sin(40x-6000t)$$

as a function of time t for x=0. Plot points every 5×10^{-4} s from t=0 to t=0.02 s.

18.3 Standing Waves and Resonance

Two sinusoidal waves which have the same amplitude, wave number, and frequency, but which travel in opposite directions, form a standing wave. The wave form does not move; peaks, for example, remain at their initial positions. Except for special points, the displacement at any point oscillates. There are, however, points where the displacement is zero for all values of the time. These points are called nodes and they are separated by $\lambda/2$ where λ is the wavelength of either of the component sinusoidal waves. To form a standing wave, the phase constants of the component waves need not be the same.

When a standing wave is on a string, all parts of the string move together with the same frequency and in phase with each other. For example, all parts of the string are at their own maximum displacement at the same time, although the maximum displacement is different for different parts of the string.

Problem 1. Use the program of Fig. 17-1 to plot

$$y(x,t) = 0.2\sin\left[20(x-350t)\right] + 0.2\sin\left[20(x+350t)\right]$$

as a function of x. Here y and x are in meters, t is in seconds. Plot points every 0.02 m from x=0 to x=0.40 m.

a. Take t=0.

b. Take $t = 2 \times 10^{-4}$ s.

Notice that the peaks and nodes remain at the same places on the string but that the displacement at any given point, except a node, changes with time.

c. Measure the distance between nodes and compare with $\lambda/2$ where λ is the wavelength of either of the traveling waves. The value of λ can be found from the functional form of the wave given above.

If the phase constant of one of the traveling waves is changed, the two waves still produce a standing wave, but the positions of the nodes and peaks are shifted.

Problem 2. Use the program of Fig. 17-1 to plot

$$y(x,t) = 0.2\sin\left[20(x-350t)+0.7\right] + 0.2\sin\left[20(x+350t)\right]$$

as a function of x. Here y and x are in meters, t is in seconds. Plot points every 0.02 m from x=0 to x=0.40 m.

a. Take t=0.

b. Take $t = 2 \times 10^{-4}$ s.

c. Verify that the distance between nodes is the same as for the standing wave of problem 1.

There is energy in a standing wave and energy is transmitted from one part of the string to neighboring parts, but no energy crosses a point where there is either a node or an antinode, a point of maximum displacement.

Problem 3. Consider the standing wave of problem 1. Take the linear mass density of the string to be 5×10^{-3} kg/m and use the program of Fig. 8-1 (Appendix C), with N=50 to answer the following.

a. For t=0, calculate the kinetic, potential, and total mechanical energy in the string between x=0 and the first node toward more positive x.

b. For $t=2\times10^{-4}$ s, calculate the kinetic, potential, and total mechanical energy in the wave between x=0 and the first node.

c. Evaluate Eqn. 17-13 for the power transmitted across the point x=0.1 m at $t=2\times10^{-4}$ s.

d. Evaluate Eqn. 17-13 for the power transmitted across a node at $t=2\times10^{-4}$ s.

Problem 4.   A string, stretched along the x axis, vibrates in a standing wave pattern. Sketch a section of the string between two nodes at a time when the displacement is positive.

a. Suppose the displacement is decreasing with time ($\partial y/\partial t<0$). Find the sign of P in Eqn. 17-13 and draw arrows to indicate the direction energy is flowing at each of two points, one on either side of the antinode.

b. Is the energy per unit length at the antinode increasing or decreasing? Is the energy per unit length at the node increasing or decreasing? Support your answers by means of an argument starting with the expressions for kinetic and potential energies in a small segment of string at an antinode or node.

c. Answer part a for the case when the displacement is increasing.

d. Answer part b for the case when the displacement is increasing.

If reflections occur at both ends of the string, it can be excited in a standing wave mode of vibration. There are certain limitations on the frequency and wavelength of the traveling waves, however. If an end is fixed, there must be a node there. If an end is free, there must be an antinode there. For a string with both ends fixed, the length must be an integer number of half wavelengths or

$$\lambda = \frac{2L}{n} , \quad n=1, 2, 3, \ldots \quad\quad (18-12)$$

Possible angular frequencies are

$$\omega = \frac{n\pi v}{L} \quad\quad (18-13)$$

327

where L is the length of the string and v is the wave speed. If one end is fixed and the other end is free, an odd integer number of quarter wavelengths must fit into the string length or

$$\lambda = \frac{4L}{2n-1} \ , \qquad n=1, \ 2, \ 3, \ \ldots .$$
(18-14)

and possible angular frequencies are

$$\omega = \frac{\pi v(2n-1)}{2L} \ .$$
(18-15)

In a sense, there are a great many traveling waves present in the string since each traveling wave forms a standing wave with its reflection from an end. When the appropriate conditions, given above, are met, all waves traveling in the same direction have the same phase at the same position and time. They are identical to each other and the string as a whole vibrates in a standing wave pattern.

Problem 5.

a. Use the trigonometric identity $\sin(A+B)=\sin(A)\cos(B)+\cos(A)\sin(B)$ to show that

$$y_m\sin(kx-\omega t) + y_m\sin(kx+\omega t) = 2y_m\sin(kx)\cos(\omega t).$$

This is a convenient form to use for the displacement when the string is vibrating as a standing wave. It clearly indicates all parts of the string vibrate in phase. The product $2y_m\sin(kx)$ is the maximum displacement at the point x and it is different for different positions along the string.

b. Consider a string which is fixed at both ends and use the above expression to show that y=0 at x=0 and at x=L provided $\lambda=2L/n$ with n an integer.

c. For the string of part b, show that the second traveling wave in the above expression is the reflection of the first at x=L and that the first traveling wave is the reflection of the second at x=0. See Eqn. 17-7. You must derive the reflection condition for x=0.

d. Consider a string which is fixed at x=0 and free at x=L and use the above expression to show that y=0 at x=0 and $y=\pm 2y_m\cos(\omega t)$ at x=L provided $\lambda=4L/(2n-1)$.

e. For the string of part d, show that the second wave in the above expression is the reflection of the first at x=L and that the first wave is the reflection of the second at x=0. See Eqn. 17-9.

The program of Fig. 2-4 (Appendix C), can also be used to plot standing waves of the form $y=2y_m\sin(kx)\cos(\omega t)$. The variable is x and its initial value is entered into X1, its final value into X2, and the increment into X3. The value of t is entered into X4 and the instruction at line 130 is

$$(2y_m)*SIN(k*X1)*COS(\omega*X4)\to X6$$

Values for y_m, k, and ω must be supplied when the machine is programmed.

The following problem is meant to help you visualize the string shape for several standing wave patterns on a string with one end fixed and one end free.

Problem 6. A 3 m string lies along the x axis with the end at x=0 fixed and the end at x=3 m free. The tension is such that the wave velocity is 17 m/s. The string carries a standing wave of the form $y=0.02\sin(kx)\cos(\omega t)$. For each of the following wavelengths, plot y as a function of x for t=0 and t=T/8, where T is the period of oscillation. Plot points every 0.25 m.
a. $\lambda=12$ m.
b. $\lambda=4$ m.
c. $\lambda=2.4$ m.

A similar demonstration can be made for a string with both ends fixed.

Problem 7. A 3 m string lies along the x axis, with both ends fixed. The wave speed is 17 m/s. The string carries a standing wave of the form $y=0.02\sin(kx)\cos(\omega t)$. For each of the following wavelengths, plot y as a function of x for t=0 and for t=T/8, where T is the period of oscillation.

a. $\lambda = 6$ m.

b. $\lambda = 3$ m.

c. $\lambda = 2$ m.

If the initial shape of the string is not one of the possible standing wave shapes, the string does not vibrate in a standing wave pattern. Instead, its displacement as a function of position and time can be described as a superposition of the standing waves which are possible for the given conditions at the string ends. The amplitudes of the various standing waves are just those required to make the displacement at t=0 equal to the initial configuration of the string.

As an example, we consider a string, fixed at both ends and started in vibration by pulling the midpoint out a distance A, then releasing it. At t=0, the string and the x axis form an equilateral triangle with altitude A. After release, the string has a displacement which is given by the superposition of all standing waves of the form

$$a_n \sin(2\pi x/\lambda_n) \cos(2\pi vt/\lambda_n) \qquad (18\text{-}16)$$

with $\lambda_n = 2L/n$, n=1, 2, 3, ... Note that this is the standing wave condition for a string with both ends fixed. The amplitudes are given by

$$a_n = \frac{8A}{(n\pi)^2} \sin(n\pi/2) \qquad (18\text{-}17)$$

Notice that waves with n even have zero amplitude and do not contribute to the total wave.

The total wave is given by the sum

$$y(x,t) = 8A \sum_{n=1}^{\infty} \frac{1}{(n\pi)^2} \sin(n\pi/2) \sin(n\pi x/L) \cos(n\pi vt/L). \qquad (18\text{-}18)$$

At t=0,

$$y(x,0) = 8A \sum_{n=1}^{\infty} \frac{1}{(n\pi)^2} \sin(n\pi/2) \sin(n\pi x/L). \qquad (18\text{-}19)$$

The following problem allows you to see how such a plucked string vibrates.

Problem 8.   A 3 m string is fixed at both ends and is plucked by pulling the
midpoint out a distance 0.02 m. It is released at t=0. Take the wave speed to
be 17 m/s.

a.   Put t=0 and approximate y by the first 5 non-vanishing terms (n=1, 3, 5, 7,
     and 9) of Eqn. 18-18. Plot y as a function of x, using points separated by
     0.25 m. Verify that the result reproduces the initial shape of the string
     to high accuracy. Higher accuracy can be obtained by using more terms in
     the series. The amplitudes, given by Eqn. 18-17, were selected so that the
     series of Eqn. 18-18 produces the initial triangular shaped displacement.
     By direct calculation, we see that they were selected correctly.

b.   Put t=T/8, where T is the period of the fundamental (n=1) oscillation.
     Plot y as a function of x, using points separated by 0.25 m. Again use the
     first 5 non-vanishing terms of Eqn. 18-18. Notice the change in the shape
     of the string. If the string were vibrating in a single standing wave, no
     change in shape would occur. Mathematically, it is the fact that the wave
     is a superposition of many standing waves that is responsible for the change
     in shape.

c.   Repeat the calculation with t=T/4 and plot the displacement of the string.
     For this value of the time, that part of the displacement attributed to the
     fundamental vibration vanishes and the graph shows the contributions of only
     the higher order terms (the overtones). Since we use only a small number of
     terms, the graph does not correctly give the shape of the string, but it does
     demonstrate the contribution of the first four overtones to the shape.

     Sometimes standing waves are created by causing one end of the string to
vibrate in simple harmonic motion. If the displacement of the end at x=0 is given
by f(t)=Acos(ωt) then, in general, the wave has the form

$$y(x,t) = \left[B\sin(kx) + C\cos(kx) \right] \cos\omega t \qquad (18\text{-}20)$$

where B and C are constants. Of course, k and ω are related by ω/k=v, the wave
speed. To meet the condition at x=0, it must be that C=A. If the end at x=L is

fixed, then Bsin(kL) + Acos(kL) = 0 or B = -(coskL/sinkL)A. The wave is then given by

$$y(x,t) = A\left[-\frac{coskL}{sinkL}\ sinkx + coskx\right]\ cos\omega t. \qquad (18-21)$$

This is a standing wave for all values of ω. All parts of the string vibrate together, in phase and at the same frequency.

When the frequency is chosen so that kL=nπ, where n is an integer, then the amplitude becomes large without bound. This phenomenon is known as resonance. In practice, of course, there are frictional losses and the amplitude remains finite.

Since k=2π/λ, the wavelengths for which resonance occurs are λ=2L/n, the same as the standing wave wavelengths when both ends are fixed. At resonance, the string is forced to vibrate at what is, in a sense, one of its natural frequencies and the string absorbs energy readily from the vibrating agent.

Eqn. 18-21 cannot be plotted when the resonance condition is met exactly, but it can be plotted for a frequency close to one of the resonance frequencies.

Problem 9.   A string is attached to a vibrator at x=0 and is fixed at x=3 m. The wave speed is 15 m/s.
a.   Show that y(x,t), as given by Eqn. 18-21, yields y=f(t) for x=0 and y=0 for x=L.
b.   Find the vibrator frequency which produces a wave with λ=1.5L and then plot y as a function of x for t=0.01 s. Take A=0.02 m and plot points for every 0.2 m.
c.   Find the vibrator frequency which produces a wave with λ=0.99L (close to a resonance) and then plot y as a function of x for t=0.01 s. Take A=0.02 m and plot points for every 0.2 m.

Chapter 19
GEOMETRICAL OPTICS: REFLECTION

The first section of this chapter gives a description of electromagnetic radiation. It is intended to be a bridge between the discussions of mechanical waves on strings in preceding chapters and discussions of optics in following chapters. This is followed by sections dealing with reflection of electromagnetic waves from plane and spherical mirrors. Computational techniques are developed so that the programmable calculator can be used to trace the rays of the reflected radiation. More detailed discussions of the physics involved can be found in Chapters 41, 42, 43, and 44 of PHYSICS, Chapters 38 and 39 of FUNDAMENTALS OF PHYSICS, and similar sections of other texts.

19.1   Electromagnetic Waves

Visible light, x-rays, infrared radiation, microwaves, and gamma rays are all electromagnetic in nature. They consist of electric and magnetic fields which are created by accelerating charge and the fields propagate outward from the charge, through a vacuum as well as through material media. In material media, some of the electromagnetic energy is absorbed and the intensity of the radiation diminishes. The speed with which the fields propagate is different for different media.

We consider radiation in empty space and in certain material media for which the propagation is similar to that in empty space. In these materials, the electric and magnetic fields produced by charges within the material are parallel, respectively, to the electric and magnetic fields of the incident radiation and the magnitudes of the fields within the material are proportional, respectively, to the magnitudes of the corresponding incident fields.

At any point in space, the electric and magnetic fields are perpendicular to each other and both are perpendicular to the direction of propagation of the wave. The wave propagates in the direction of the vector cross product $\vec{E} \times \vec{B}$.

In many instances, we consider sinusoidal plane waves, for which the fields at any point in space oscillate sinusoidally in time and, at any time, they vary sinusoidally in space. If the wave is moving in the positive z direction, the fields might be given by

$$\vec{E} = \hat{i}E_0 \sin(kz-\omega t) \qquad (19-1)$$

and

$$\vec{B} = \hat{j}B_0 \sin(kz-\omega t) \qquad (19-2)$$

respectively. Here $k=2\pi/\lambda$ and $\omega=2\pi/T$ where λ is the wavelength and T is the period.

The speed v of the wave is related to these quantities by

$$v = \lambda/T = \omega/k. \qquad (19-3)$$

In a vacuum, the speed is denoted by c and has the value 2.997925×10^8 m/s, while in a material medium, the speed has another value, dependent on the medium. Rather than give the speed in the material, it is usual to give the index of refraction n for the material. The speed in the material is then related to the speed in vacuum by

$$v = c/n. \qquad (19-4)$$

For air, n is slightly greater than 1 but, for our purposes, n=1 is sufficiently accurate. For most glasses, n is in the range from 1.5 to 2 for optical frequencies.

Notice that the expressions for the fields have the form of traveling waves. The electric field at some point in space and at some time is the same as the electric field at a point with less positive z coordinate, but at an earlier time. The field vector travels at speed v and the field at z,t is the same as the field at $z-\Delta z$, $t-\Delta z/v$. The same statement can be made about the magnetic field.

The two fields are in phase. At any point in space the electric and magnetic fields reach their respective maxima at the same time and they both vanish at the same time.

The direction of the fields are related in such a way that the cross product $\vec{E} \times \vec{B}$ is in the direction of propagation. When E_x changes sign, so does B_y and the direction of propagation remains the same. To describe fields which propagate in the negative z direction, not only must the sign in front of ωt be changed in Eqns. 19-1 and 19-2, but the direction of one of the fields must also be changed.

Eqns. 19-1 and 19-2 represent fields which exist throughout space, not just

along the z axis. For these waves, the electric fields at two points with the same z
coordinate are exactly the same. For these waves, the z axis and lines parallel to it
are called rays. Rays give the direction of travel of the waves and the electric fields
at all points on a plane perpendicular to a ray are in phase, as are the magnetic fields.
Surfaces of uniform phase are called wave fronts.

We shall also deal with spherical waves. These waves emanate from point sources
and move along rays which are directed radially outward from the source. The surfaces
of uniform phase are spheres centered at the source and are, again, everywhere
perpendicular to the rays from the source. Mathematically, any component of either
field can be described by an expression of the form

$$(A/r) \sin(kr - \omega t) \qquad\qquad (19\text{--}5)$$

where r is the distance from the source to the field point. The electric and magnetic
fields are perpendicular to each other and to the direction of propagation.

The spherical waves described above are said to be divergent since they diverge
from a point source. We shall also need to consider convergent spherical waves, which
move radially inward toward a point. Mathematically, these are described by an
equation which is the same as Eqn. 19-5 except that the sign in front of ωt is changed.

A great many of the applications of optics have to do with the formation of
images. Spherical waves diverge from a point source and a system of lenses is designed
to convert parts of them to waves which converge on a point, the image of the source.

One way to visualize the propagation of waves is to think of the wave fronts as
moving with speed v along the rays. Pick a value for the phase and imagine a surface
over which the wave has that phase at some particular time. The surface is a plane for
plane waves and a sphere for spherical waves. Rays penetrate the surface at right
angles. A short time Δt later, the surface corresponding to the selected phase is at a
different position. It has moved a distance $v\Delta t$ along the rays.

In this and the next chapter, we consider what is called geometrical or ray
optics. More specifically, we deal only with the rays and their changes in direction
when the radiation strikes the boundary between two different materials. We trace rays
from the source to the boundary and then trace the rays of the reflected and
transmitted radiation. In so doing we ignore those phenomena which arise from the wave

336

nature of the radiation. Wave phenomena will be discussed in a later chapter.

19.2   Reflection at a Plane Surface

When a radiation wave, moving in one medium, is incident on a boundary with
another medium, it is partially reflected back into the first medium and partially
transmitted into the second. We concentrate here on the reflected wave. We follow a
ray of the incident wave to the boundary and find its intersection with the boundary.
We then follow a ray of the reflected wave as it leaves the boundary at the point of
intersection we found. The situation is depicted in Fig. 19-1. The boundary is a
plane which is perpendicular to the x,y plane. The edge is shown in the diagram. The
incident ray lies in the x,y plane and is represented by the vector $\vec{r}_1$. The reflected
ray also lies in the x,y plane and it is represented by the vector $\vec{r}_2$. $\hat{s}$ is a unit
vector perpendicular to the surface and it points into the medium in which the incident
wave propagates. θ_1 is called the angle of incidence, θ_2 is called the angle of
reflection, and these two angles are equal. The law of reflection is expressed by

$$\theta_1 = \theta_2. \tag{19-6}$$

The incident radiation may be plane or spherical in nature. In either case,

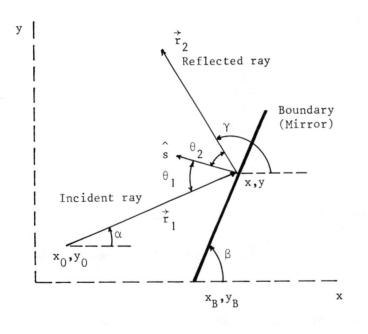

Figure 19-1.
Reflection at a plane
boundary.

each pair of rays, consisting of an incident and reflected ray, obeys the law of
reflection. If plane waves are incident on the boundary, then the incident ray is part
of a system of rays, all of which are parallel to each other and each of which have
associated reflected waves. The reflected rays are also parallel to each other and the
reflected radiation also consists of plane waves.

If the radiation emanates from a point source, the ray shown is part of a
system of rays which diverge from that source. They strike the boundary with different
angles of incidence and give rise to a system of reflected rays which, if extended,
diverge from a point on the right side of the boundary.

We now develop a calculator program to find the point of intersection of the
incident ray and the boundary and to find the direction of the reflected ray. The
incident ray is specified by giving one point on the ray, with coordinates x_0, y_0, and the
angle α between the ray and the x axis. The point might be the source of the radiation
or any other point on the ray. Likewise, the boundary is specified by giving one point,
with coordinates x_B, y_B, and the angle β between the boundary and the x axis. The angles
α and β increase in the counterclockwise direction when viewed as in the diagram. If
$\alpha = 180^\circ$, the ray points to the left along the x axis and if $\beta = 180^\circ$, the boundary is
parallel to the x axis with the reflecting surface facing in the negative y direction.

If the point of intersection is a distance ℓ from x_0, y_0 and a distance d from
x_B, y_B, then the x coordinate of the point of intersection is given by

$$x = x_0 + \ell\cos\alpha \qquad\qquad (19\text{-}7)$$
and also by
$$x = x_B + d\cos\beta. \qquad\qquad (19\text{-}8)$$

The y coordinate of the point of intersection is given by

$$y = y_0 + \ell\sin\alpha \qquad\qquad (19\text{-}9)$$
and by
$$y = y_B + d\sin\beta. \qquad\qquad (19\text{-}10)$$

If the two expressions for x are equated to each other and the two expressions for y
are also equated to each other, the results are

$$x_0 + \ell\cos\alpha = x_B + d\cos\beta \qquad\qquad (19\text{-}11)$$
and
$$y_0 + \ell\sin\alpha = y_B + d\sin\beta. \qquad\qquad (19\text{-}12)$$

338

These are two equations for the two unknowns, ℓ and d. Elimination of d and solution for ℓ produces

$$\ell = \frac{(x_B - x_0)\sin\beta + (y_0 - y_B)\cos\beta}{\sin(\beta - \alpha)} \qquad (19\text{-}13)$$

and substitution of this expression into Eqns. 19-7 and 19-9 yields

$$x = x_0 + \frac{(x_B - x_0)\sin\beta - (y_B - y_0)\cos\beta}{\sin(\beta - \alpha)}\cos\alpha \qquad (19\text{-}14)$$

and

$$y = y_0 + \frac{(x_B - x_0)\sin\beta - (y_B - y_0)\cos\beta}{\sin(\beta - \alpha)}\sin\alpha \qquad (19\text{-}15)$$

respectively. Here the trigonometric identity $\sin\alpha\cos\beta - \cos\alpha\sin\beta = \sin(\alpha - \beta)$ was used.

For the incident ray to intersect the boundary on the reflecting side, the condition $\hat{r}_1 \cdot \hat{s} < 0$ must hold. Otherwise, the ray is parallel to the surface or is directed away from it. Since

$$\hat{r}_1 = \cos\alpha\ \hat{i} + \sin\alpha\ \hat{j} \qquad (19\text{-}16)$$

and

$$\hat{s} = -\sin\beta\ \hat{i} + \cos\beta\ \hat{j}, \qquad (19\text{-}17)$$

the condition can be written

$$\sin(\beta - \alpha) > 0. \qquad (19\text{-}18)$$

The coordinates of the intersection point have now been found. We go on to find the direction of the reflected ray. It is specified by the angle γ between the ray and the x axis; the angle increases in the counterclockwise direction when viewed as in Fig. 19-1.

If $\hat{r}_2$ is a unit vector along the reflected ray, then its component in a direction perpendicular to $\hat{s}$ is the same as the similar component of $\hat{r}_1$ and its component along $\hat{s}$ has the same magnitude as the similar component of $\hat{r}_1$ but has the opposite sign. Since the component of $\hat{r}_1$ along $\hat{s}$ is

$$\hat{r}_1 \cdot \hat{s} = -\cos\theta_1 = -\sin(\beta-\alpha), \qquad\qquad (19\text{-}19)$$

$\hat{r}_2$ is given by

$$\hat{r}_2 = \hat{r}_1 + 2\cos\theta_1 \; \hat{s} = \hat{r}_1 + 2\sin(\beta-\alpha) \; \hat{s}$$

$$= \left[\cos\alpha - 2\sin(\beta-\alpha)\sin\beta\right]\hat{i} + \left[\sin\alpha + 2\sin(\beta-\alpha)\cos\beta\right]\hat{j} \qquad (19\text{-}20)$$

where we have used Eqn. 19-16 to substitute for $\hat{r}_1$ and Eqn. 19-17 to substitute for $\hat{s}$. Notice that twice the component of $\hat{r}_1$ along $\hat{s}$ must be subtracted from $\hat{r}_1$.

The tangent of γ is the ratio of the y component of this vector to the x component and

$$\gamma = \tan^{-1} \frac{\sin\alpha + 2\sin(\beta-\alpha)\cos\beta}{\cos\alpha - 2\sin(\beta-\alpha)\sin\beta} . \qquad\qquad (19\text{-}21)$$

There are two values of γ which satisfy this equation and they differ by 180°. The choice of the correct value can be made by comparing the cosine of the angle produced by the machine with $\cos\alpha - 2\sin(\beta-\alpha)\sin\beta$. If they have opposite signs, 180° is added to the angle to obtain γ.

There is a computational problem if $\cos\alpha - 2\sin(\beta-\alpha)\sin\beta = 0$. Then γ is either 90° or 270° but many machines cannot correctly evaluate Eqn. 19-21 in that event. When the denominator which appears in the equation vanishes, the numerator is evaluated. It is +1 for $\gamma=90^\circ$ and -1 for $\gamma=270^\circ$.

A flow chart is shown in Fig. 19-2. The program is written for angles in degrees. Information about the incident ray is entered at line 100: x_0 in X1, y_0 in X2, and α in X3. At line 110, information about the reflecting surface is entered: x_B in X4, y_B in X5, and β in X6. At line 120, $\sin(\beta-\alpha)$ is calculated and stored in X7. It is checked, at line 130, to see if it is positive. If it is negative or zero, no reflection occurs and the machine proceeds to line 300 where the number 1×10^{60} is displayed as a signal. After restarting, the machine goes to line 110 to consider the same ray incident on another boundary.

If $\sin(\beta-\alpha)$ is positive, the machine goes to line 140, where ℓ is calculated, using Eqn. 19-13. Then, at line 150, x and y are calculated, using Eqns. 19-7 and 19-9 respectively. Results are displayed at line 160 and x and y are stored in X1 and

340

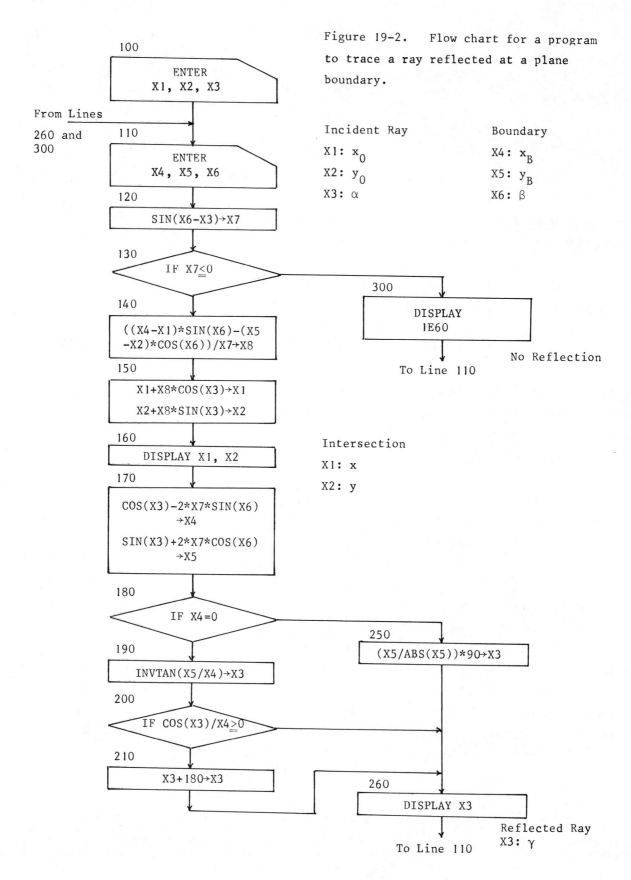

Figure 19-2. Flow chart for a program to trace a ray reflected at a plane boundary.

100
ENTER
X1, X2, X3

From Lines
260 and
300

110
ENTER
X4, X5, X6

120
SIN(X6-X3)→X7

130
IF X7≤0

140
((X4-X1)*SIN(X6)-(X5
-X2)*COS(X6))/X7→X8

150
X1+X8*COS(X3)→X1
X2+X8*SIN(X3)→X2

160
DISPLAY X1, X2

170
COS(X3)-2*X7*SIN(X6)
→X4
SIN(X3)+2*X7*COS(X6)
→X5

180
IF X4=0

190
INVTAN(X5/X4)→X3

200
IF COS(X3)/X4≥0

210
X3+180→X3

250
(X5/ABS(X5))*90→X3

260
DISPLAY X3

300
DISPLAY
1E60

To Line 110

No Reflection

To Line 110

Reflected Ray
X3: γ

Incident Ray
X1: x_0
X2: y_0
X3: α

Boundary
X4: x_B
X5: y_B
X6: β

Intersection
X1: x
X2: y

X2 respectively. They give a point on the reflected ray and are retained to be used if that ray is incident on another boundary.

The calculation of γ now begins. At line 170, the numerator and denominator which appear in Eqn. 19-21 are computed separately and stored in X5 and X4 respectively. Since x_B and y_B are no longer needed, their erasure causes no harm. If the denominator vanishes, γ must be 90° or 270° and the machine, at line 250, multiplies 90 by the numerator divided by the magnitude of the numerator. This produces 90 if the numerator is positive and -90 if the numerator is negative. -90° is, of course, equivalent to 270°. The value is stored in X3 and displayed at line 260.

If, at line 180, the denominator does not vanish, the machine proceeds to line 190, where Eqn. 19-21 is evaluated and the result placed in X3. This is either γ or $\gamma-180^\circ$. At line 200, the signs of the cosine of the angle found and the denominator of Eqn. 19-21 are compared. If they are the same, X3 contains γ and the machine goes to line 260 where the value is displayed. If they are not the same, the machine goes to line 210 where 180 is added to the angle in X3 before the result is displayed.

In any event, γ is placed in X3, where it is ready to be used if the reflected ray is incident on a second boundary.

After displaying γ, the machine goes to line 110 where information about another boundary is entered. In this way multiple reflections can be considered. The reflected ray becomes the incident ray for the next surface. Appropriate values for x_0, y_0, and α are already stored in X1, X2, and X3 respectively.

If a ray is to be plotted, it is convenient to know the coordinates of two points on the ray. For the incident ray two points, at x_0,y_0 and at x,y, are known and the ray can be plotted by drawing the straight line which joins them. For the reflected ray, however, only the point of intersection with the last surface is known. If another surface is inserted into the problem, the machine can be used to produce the coordinates of another point on the reflected ray. A choice for the inserted boundary can be made once the direction of the last ray has been determined. Remember that β must be chosen so that $\hat{s}$ points into the region of the incident ray.

The first few problems can be used to test the program. All of the problem descriptions make use of the angle conventions described above and depicted in Fig. 19-1.

Problem 1.　A mirror passes through the point x=0.05 m, y=0 and makes an angle of $\beta=45^{\circ}$ with the x axis. It reflects rays incident from the left. All incident rays pass through the origin. For each of the ray directions given below, tell if reflection occurs and, if it does, find the point on the mirror where it does and find the direction of the reflected ray. On a piece of graph paper, draw the mirror surface, the normal to the surface, the incident rays, and the reflected rays. Use a protractor to verify, in each case, that the angle of incidence is equal to the angle of reflection.

a.　$\alpha=-45^{\circ}$.

b.　$\alpha=-20^{\circ}$.

c.　$\alpha=0$.

d.　$\alpha=+20^{\circ}$.

e.　$\alpha=+45^{\circ}$.

Problem 2.　A corner reflector consists of a mirror surface through x=0.08 m, y=0, with its reflecting surface facing toward negative x and a second mirror surface through x=0, y=0.08 m, with its reflecting surface facing toward negative y. A light ray which is reflected from both mirrors exits parallel to the incident ray. Trace the following rays. You must pay attention to the order in which the two surfaces are intersected by the rays.

a.　$x_0=-0.03$ m, $y_0=0$, $\alpha=60^{\circ}$.

b.　$x_0=-0.03$ m, $y_0=0$, $\alpha=45^{\circ}$.

c.　$x_0=+0.01$ m, $y_0=0$, $\alpha=20^{\circ}$.

d.　$x_0=+0.05$ m, $y_0=0$, $\alpha=35^{\circ}$.

Problem 3.　A mirror surface passes through x=0.05 m, y=0 and makes an angle of 55° with the x axis. A second surface passes through x=0.08 m, y=0.07 m and makes an angle of 100° with the x axis. A light ray, incident on the first mirror, passes through the point x=0, y=-0.01 m and makes an angle of 30° with the x axis.

a.　Trace the ray on a graph and find its direction after the second reflection.

b.　Analytically find the orientation of a third surface so that the ray, after reflection from the three surfaces, is parallel to the direction of incidence. Place the third surface at any convenient location.

c.　Do other rays emerge parallel to the direction they enter the mirror system consisting of the three mirrors? Try $x_0=0$, $y_0=-0.01$ m, $\alpha=20^{\circ}$ and trace the ray through three reflections. You may need to move the third mirror, but do not change its orientation.

The rays associated with spherical waves, upon reflection from a plane surface, all diverge from the same point, located behind the surface. The reflected waves are all like those that would be produced if there were a point source at that point and there were no mirror surface. The point is an image of the true source and, since there is in reality no radiation emanating from it, it is called a virtual image.

Problem 4.   A point source of spherical waves is located at the origin. Some of the waves are reflected from a surface which passes through x=0.05 m, y=0 and makes an angle of 72° with the x axis.

a.   Trace representative rays from the source to the surface and plot them and their reflections. Start with α=-80° and increment it by 30° for each new ray considered. Continue until you obtain a ray which is not reflected.

b.   Extend the reflected rays backward and verify that, to within drawing error, they meet at a single point. To obtain more precision, you can use the program of Fig. 19-2 to find the coordinates of the intersection point of the rays. Treat one of the rays as if it were a reflecting surface.

c.   Draw the line joining the source and image and verify that it is perpendicular to the reflecting surface and that the source-to-mirror distance is the same as the image-to-mirror distance.

Problem 5.   Consider the corner reflector described in problem 2. A point source, located at the origin, emits spherical waves. Trace rays from the source to the reflector and then trace their reflections. Take α=-30°, -15°, 0, 30°, 60°, 90°, and 105°. Graphically find all points of intersection of the reflected rays, extended backward into the region behind the reflector.

Notice that there are three images. Rays which are reflected once, from one mirror, diverge from one image, rays which are reflected once, from the other mirror, diverge from a second image, and rays which are reflected twice diverge from the third image. Before the second reflection, these last rays diverge from one or the other of the first two images. Thus the third image is, in a sense, the image of the other two. That is, it may be considered to be the image formed by the second mirror of the image in the first mirror.

19.3   Reflection at a Spherical Surface

We consider a ray which is incident on a spherical reflecting surface and desire to find the point of intersection and to find the direction of the reflected ray. As is true for reflection at a plane surface, the incident and reflected rays make the same angle with the outward normal to the surface at the point of reflection.

The geometry is shown in Fig. 19-3 for a convex surface and in Fig. 19-4 for a concave surface. The incident ray and the center of the sphere define a plane, which we take to be the x,y plane with the x axis passing through the sphere center. The incident ray passes through the point x_0, y_0 and makes the angle α with the positive x direction. $\vec{r}_1$ is a vector along the incident ray from x_0, y_0 to the point of intersection, $\vec{r}_2$ is a vector along the reflected ray, $\hat{s}$ is a unit vector normal to the surface at the point of intersection, and $\vec{R}$ is a vector from the sphere center to the point of intersection. $\vec{r}_2$ makes the angle γ with the positive x direction.

It is usual in dealing with spherical surfaces to specify the radius R of the sphere and the point, called the vertex, where the sphere cuts the x axis. The coordinate of the vertex is denoted by x_v and, by convention, R is positive for concave surfaces and negative for convex surfaces. The sphere center is at $(x_v - R)\hat{i}$ in either case. For both types of surface $\hat{s} = -\vec{R}/R$, where the sign convention for R is used.

Figure 19-3.   Reflection from a convex spherical surface. R is negative.

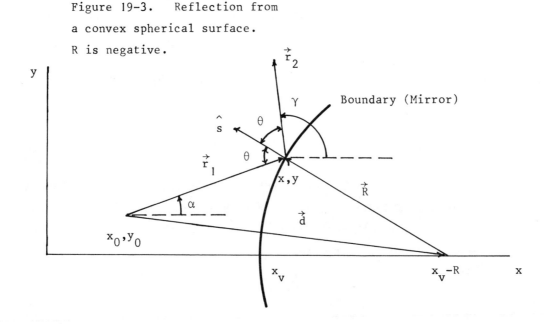

Figure 19-4. Reflection from a concave spherical surface. R is positive.

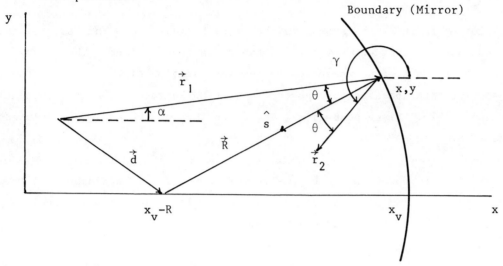

The first job is to find the intersection point x,y of the incident ray and the surface. The equation of the circle formed by the sphere where it cuts the x,y plane is

$$(x-x_v+R)^2 + y^2 = R^2 \tag{19-22}$$

for both convex and concave surfaces. If x_0, y_0 is a distance ℓ from the point of intersection, $\vec{r}_1 = \ell\cos\alpha\,\hat{i} + \ell\sin\alpha\,\hat{j}$,

$$x = x_0 + \ell\cos\alpha, \tag{19-23}$$

and
$$y = y_0 + \ell\sin\alpha. \tag{19-24}$$

Substitution of these expressions into Eqn. 19-22 yields

$$\ell^2 - 2\ell\left[(x_v-R-x_0)\cos\alpha - y_0\sin\alpha\right] + (x_v-R-x_0)^2 + y_0^2 - R^2 = 0 \tag{19-25}$$

after some algebraic manipulation and use of the identity $\cos^2\alpha + \sin^2\alpha = 1$.

This equation has the two solutions

$$\ell = \left[(x_v - R - x_0)\cos\alpha - y_0\sin\alpha\right]$$

$$\pm\left\{\left[(x_v - R - x_0)\cos\alpha - y_0\sin\alpha\right]^2 - (x_v - R - x_0)^2 - y_0^2 + R^2\right\}^{\frac{1}{2}}. \qquad (19\text{-}26)$$

The correct solution is chosen by requiring that $\vec{r}_1 \cdot \hat{s}$ be negative. This assures that ℓ is positive and that the incident ray strikes the reflecting surface. The other solution, when it is positive, corresponds to the intersection of the ray with the far side of the sphere, after passing through the reflecting surface. A negative solution for ℓ is not admissible. Since $\cos\theta = -\hat{r}_1 \cdot \hat{s}$, the criterion for the choice of solution is the same as $\cos\theta > 0$.

We now calculate $\cos\theta$. Let $\vec{d}$ be the vector from x_0, y_0 to the center of the sphere. Then $\vec{r}_1 = \vec{d} + \vec{R}$. This is solved for $\vec{R}$ and the result is substituted into $\hat{s} = -\vec{R}/R$ to give $\hat{s} = -(\vec{r}_1 - \vec{d})/R$. Take the scalar product with $\hat{r}_1$ and use $\vec{r}_1 = \ell\hat{r}_1$ to find $\cos\theta = (\ell - \hat{r}_1 \cdot \vec{d})/R$.

Now $\hat{r}_1 = \hat{i}\cos\alpha + \hat{j}\sin\alpha$ and $\vec{d} = (x_v - R - x_0)\hat{i} - y_0\hat{j}$, so $\cos\theta = \left[\ell - (x_v - R - x_0)\cos\alpha + y_0\sin\alpha\right]/R$. Finally, substitute the value for ℓ from Eqn. 19-26 to find

$$\cos\theta = \pm(1/R)\left\{\left[(x_v - R - x_0)\cos\alpha - y_0\sin\alpha\right]^2 - (x_v - R - x_0)^2 - y_0^2 + R^2\right\}^{\frac{1}{2}}. \qquad (19\text{-}27)$$

Since R is negative for convex surfaces, we select the negative sign in Eqns. 19-26 and 19-27 when dealing with those surfaces. For concave surfaces, we select the positive sign. In either case, we use $R/|R|$ to determine the correct sign and write

$$\cos\theta = (1/|R|)\left\{\left[(x_v - R - x_0)\cos\alpha - y_0\sin\alpha\right]^2 - (x_v - R - x_0)^2 - y_0^2 + R^2\right\}^{\frac{1}{2}}. \qquad (19\text{-}28)$$

Then Eqn. 19-26 becomes

$$\ell = (x_v - R - x_0)\cos\alpha - y_0\sin\alpha + R\cos\theta. \qquad (19\text{-}29)$$

Once ℓ is found, the coordinates of the intersection point are calculated using Eqns. 19-23 and 19-24.

Just as for reflection at a plane surface, the unit vector in the direction of the reflected ray is given by

$$\hat{r}_2 = \hat{r}_1 + 2\hat{s}\cos\theta. \qquad (19\text{-}30)$$

The x component of $\hat{r}_2$ is

$$(\hat{r}_2)_x = \cos\alpha - 2\cos\theta\left[\ell\cos\alpha - (x_v-R-x_0)\right]/R$$

$$= \cos\alpha - 2(x-x_v+R)\cos\theta/R \qquad (19\text{-}31)$$

and the y component is

$$(\hat{r}_2)_y = \sin\alpha - 2\cos\theta\left[\ell\sin\alpha + y_0\right]/R$$

$$= \sin\alpha - 2y\cos\theta/R. \qquad (19\text{-}32)$$

Here we have used $\hat{s}=-(\vec{r}_1-\vec{d})/R$ and substituted from Eqns. 19-23 and 19-24.

The ratio of the y to the x component is the tangent of γ and γ can be found using

$$\gamma = \tan^{-1}\frac{\sin\alpha - 2y\cos\theta/R}{\cos\alpha - 2(x-x_v+R)\cos\theta/R}. \qquad (19\text{-}33)$$

If the denominator vanishes, γ is either $+90^o$ or -90°, according to the sign of the numerator. In general, there are two values of γ which satisfy Eqn. 19-33. If the cosine of the angle produced by the calculator does not agree in sign with the denominator of Eqn. 19-33, we add 180° to it.

The flow chart is shown in Fig. 19-5. At line 100, x_0, y_0, and α are entered and stored in X1, X2, and X3, respectively. These numbers describe the incident ray. Recall that α is positive in the counterclockwise direction when viewed as in the diagrams. Also entered, at line 110, are the coordinate x_v of the vertex and the radius R of the sphere. Remember that R is negative for convex surfaces and positive for concave surfaces.

At line 120, the combination x_v-R-x_0 is calculated and stored in X6, the combination $(x_v-R-x_0)\cos\alpha - y_0\sin\alpha$ is calculated and stored in X7, and the combination $\left[(x_v-R-x_0)\cos\alpha - y_0\sin\alpha\right]^2 - (x_v-R-x_0)^2 - y_0^2 + R^2$ is calculated and stored in X8. The sign of the last quantity is tested at line 130. If it is negative, the ray misses the surface and the machine goes to line 300, where the number 1×10^{60} is displayed to signal that there is no intersection of the ray and the surface.

348

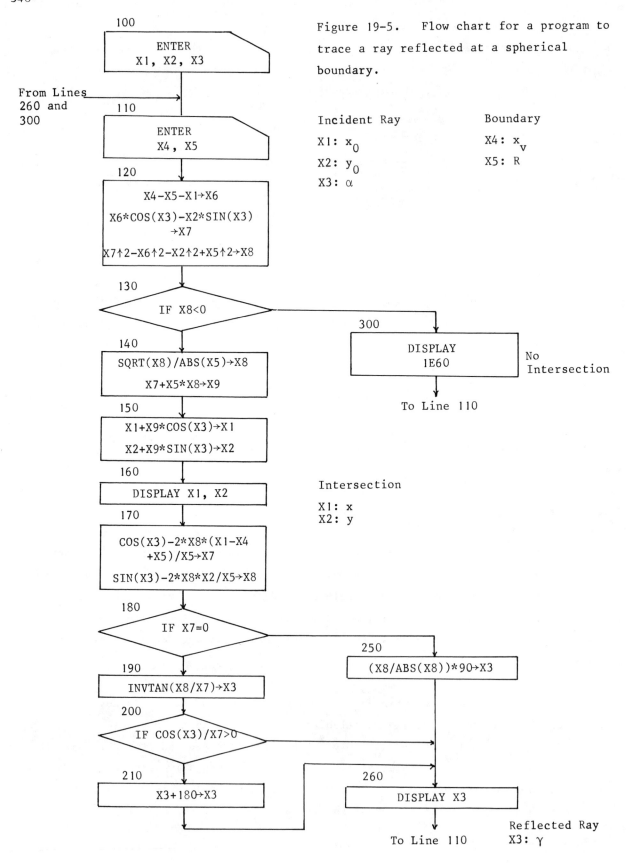

Figure 19-5. Flow chart for a program to trace a ray reflected at a spherical boundary.

From Lines 260 and 300

100 ENTER X1, X2, X3

110 ENTER X4, X5

120
X4−X5−X1→X6
X6*COS(X3)−X2*SIN(X3)→X7
X7↑2−X6↑2−X2↑2+X5↑2→X8

130 IF X8<0

140
SQRT(X8)/ABS(X5)→X8
X7+X5*X8→X9

150
X1+X9*COS(X3)→X1
X2+X9*SIN(X3)→X2

160 DISPLAY X1, X2

170
COS(X3)−2*X8*(X1−X4+X5)/X5→X7
SIN(X3)−2*X8*X2/X5→X8

180 IF X7=0

190 INVTAN(X8/X7)→X3

200 IF COS(X3)/X7>0

210 X3+180→X3

250 (X8/ABS(X8))*90→X3

260 DISPLAY X3

300 DISPLAY 1E60

No Intersection

To Line 110

Intersection
X1: x
X2: y

To Line 110

Reflected Ray
X3: γ

Incident Ray
X1: x_0
X2: y_0
X3: α

Boundary
X4: x_v
X5: R

At line 140, Eqn. 19-28 is used to compute $\cos\theta$ and Eqn. 19-29 is used to compute ℓ. These quantities are stored in X8 and X9 respectively. Note that the number previously stored in X8 is not needed once $\cos\theta$ is computed.

The coordinates of the intersection are calculated at line 150. Eqns. 19-23 and 19-24 are used, x is stored in X1, and y is stored in X2. These numbers are then displayed.

At line 170, the x and y components of $\hat{r}_2$ are calculated and stored in X7 and X8 respectively. Eqns. 19-31 and 19-32 are used. If the x component vanishes, the machine goes to line 250, where γ is set equal to 90° (if the y component is positive) or to -90° (if the y component is negative). If the x component does not vanish, the machine goes to line 190, where the arc tangent of the ratio of the components is evaluated. Then the cosine of the resulting angle is compared in sign to the x component of $\hat{r}_2$. If they are the same, X3 contains γ and nothing more need be done. If they are different, 180° is added to X3 to produce γ. In any event, the value of γ is stored in X3 and displayed at line 260.

The machine returns to line 110 to consider a new surface. The reflected ray for the first surface becomes the incident ray for the second surface. Information about the ray has already been stored in appropriate memory locations.

Even when it is not given in the problem, a second surface can be inserted for the sole purpose of finding its intersection with the reflected ray. Then two points on that ray are known and the ray can be traced by drawing the line joining them. The position of the vertex and the magnitude of the sphere radius must be chosen so that the ray intersects the inserted surface. The sign of the radius is usually immaterial.

Sometimes it is desired to find the intersection of an emerging ray with the x axis. If the x coordinate of that point is denoted by x_i, then $x_i = x - y/\tan\gamma$ where x and y are the coordinates of the intersection with the last surface. This calculation can be incorporated into the program after line 260 or it can be carried out separately. If $\tan\gamma = 0$, the machine will signal that an improper arithmetic operation has been attempted but then the ray is parallel to the x axis.

The first two problems constitute tests of the program.

350

Problem 1. A convex spherical reflecting surface has a 0.07 m radius and is situated with its vertex at x_v=0.05 m and its center on the x axis. A source of spherical light waves is at the origin.

a. Draw a diagram, to scale, showing the surface and the source.

b. For each of the following values of α, use the program of Fig. 19-5 to find if the ray intersects the surface and, if it does, trace the ray and its reflection. Take $\alpha=-40^\circ$, -30°, -20°, 0°, 20°, 30°, and 40°.

c. At each of the intersection points, draw the normal vector (it is a continuation of the line from the intersection point to the sphere center) and use a protractor to verify that the angles of incidence and reflection are the same.

Save the values you obtain for the coordinates of the intersection points and for γ. They will be used in a later problem.

Problem 2. A concave spherical reflecting surface has a 0.07 m radius and is situated so that its vertex is at x_v=0.05 m and its center is on the x axis. A point source is located at the origin and emits spherical light waves.

a. Draw a diagram, to scale, showing the surface and the source.

b. For each of the following values of α, use the program of Fig. 19-5 to trace the ray and its reflection. Take $\alpha=-40^\circ$, -30°, -20°, 0°, 20°, 30°, and 40°.

c. At each of the intersection points, draw the normal vector and use a protractor to verify that the angles of incidence and reflection are the same.

Save the values you obtain for the coordinates of the intersection points and for γ. They will be used in a later problem.

The next few problems deal with rays that are never far from the optic axis. These rays are said to be paraxial and, if they emanate from a point source, the reflected rays either diverge from nearly the same point behind the surface or converge to nearly the same point in front of the surface. In the first case, the rays are said to form a virtual image of the source while, in the second case,

they are said to form a real image of the source. In practice, the radiation
reaching the surface can be limited to that with paraxial rays by placing an
opaque sheet with a small opening in front of the surface.

For paraxial rays, there is a simple relationship between the positions
of the object and image. If o is the difference in the x coordinates of the
object and vertex and i is the difference in the x coordinates of the image and
vertex, then

$$\frac{1}{o} + \frac{1}{i} = \frac{2}{R} \; .$$

(19-34)

There is a sign convention associated with this equation. The image distance i
is positive if the image is real (in front of the surface) and negative if the
image is virtual (behind the surface). The object distance o is positive for
a source in front of the surface. As before, R is negative for convex surfaces
and positive for concave surfaces. Eqn. 19-34 is an approximation, good for
paraxial incident rays, and is more exact the smaller α is.

Problem 3.   Consider the surface and source of problem 1.
a.   Use the program of Fig. 19-5 to find the intersection points and the
direction of the reflected rays for incident rays with $\alpha=-0.5^{o}$ and $+0.5^{o}$
respectively.
b.   Use the program of Fig. 19-2 to find the intersection of the two reflected
rays, extended into the region behind the surface. Consider one of the
rays to be part of a plane mirror and the other to be a ray incident on
it. You may have to change one of the angles by 180° for the program to
work. This intersection point is the position of the virtual image.
c.   To check that other reflected rays diverge from the same point, repeat the
calculation of parts a and b for incident rays with $\alpha=-1^{o}$ and $+1^{o}$
respectively.
d.   Use the result of part b to find the image distance i. Also calculate i
using Eqn. 19-34 and compare answers. Agreement is not exact.

Save your results for later use.

If the reflecting surface is concave and the source is more than R/2 from the vertex, then a real image is formed by paraxial rays. The reflected rays now converge toward a common point in front of the surface.

Problem 4.   Consider the surface and source of problem 2.

a.   Use the program of Fig. 19-5 to find the intersection points and the directions of the reflected rays for incident rays with $\alpha = -0.5^o$ and $+0.5^o$ respectively.

b.   Use the program of Fig. 19-2 to find the intersection of the two reflected rays. See part b of problem 3.

c.   Check to see if other paraxial rays converge to the same point. Repeat the calculation of parts a and b for incident rays with $\alpha = -1^o$ and $+1^o$ respectively.

d.   Use the result of part b to find the image distance i. Also calculate i using Eqn. 19-34 and compare answers.

Save your results for later use.

If the reflecting surface is concave and the source is less than R/2 from the vertex, a virtual image is formed by the paraxial rays.

Problem 5.   A point source is 0.03 m from the vertex of a concave reflecting surface with a 0.08 m radius.

a.   Use the program of Fig. 19-5 to find the intersection points and the directions of the reflected rays for incident rays with $\alpha = -0.5^o$ and $+0.5^o$ respectively.

b.   Use the program of Fig. 19-2 to find the intersection of the reflected rays, extended to the region behind the surface. See part b of problem 3.

c.   Check to see if other paraxial rays diverge from the same point. Repeat the calculation of parts a and b for incident rays with $\alpha = -1^o$ and $+1^o$ respectively.

d.   Use the results of part b to find the image distance i. Also calculate i using Eqn. 19-34 and compare answers.

Save your results for later use.

If the source is moved off the optic axis by changing its y coordinate but not its x coordinate, the paraxial rays form an image and the image of the displaced source has nearly the same x coordinate as the image of the undisplaced source. Since o and R are the same, Eqn. 19-34 produces the same value of i for the two situations.

Problem 6.   Consider the spherical reflecting surface of problem 1 and place the source at x=0, y=0.002 m.

a.   Find the intersection points and the directions of the reflected rays for incident rays with $\alpha=-0.5^o$ and $\alpha=+0.5^o$ respectively. Use the program of Fig. 19-2 to find the intersection of the two reflected rays. The x coordinate of this point should be nearly the same as the x coordinate of the image in problem 3.

b.   Repeat the calculation of part a for incident rays with $\alpha=-1^o$ and $\alpha=+1^o$ respectively.

c.   Compare the answers to parts a and b to each other and to the answer to part b of problem 3.

The image of an extended body can be found by finding the images of representative points on the body, each point being considered a source of spherical waves. We consider a line source, parallel to the y axis. In order to distinguish the two ends of the line, it is convenient to take the source to be an arrow pointing in the positive y direction with its tail resting on the x axis.

As an example, consider the reflecting surface of problem 1 and suppose a 0.002 m arrow is located at x=0. The length of the arrow is small enough that rays with small α can be considered paraxial to a reasonable approximation and large enough that you can see some failure of the paraxial approximation. According to the results of problems 3 and 6, the image of the tail is at x=0.0706 m, y=0 and the image of the head is at x=0.0706 m, $y=8.2\times10^{-4}$ m. The image is nearly a straight line which extends in the positive y direction. It is shorter than the object. The surface is said to have a magnification factor of 0.41, a number which is the ratio of the image length to the object length. The magnification is taken to be positive if the image is erect (in the same direction as the object) and negative if the image is inverted. For paraxial rays, the magnification is given by $-i/o$.

Problem 7.

a. For each of the reflecting surfaces of problems 2 and 5, find the image of an arrow with its tail at x=0, y=0 and its head at x=0, y=0.002 m. Use rays with $\alpha=-0.5^{o}$ and $+0.5^{o}$ respectively. Also use the results of problems 4 and 5.

b. For each of these surfaces, find the magnification. In each case, compare the magnification to $-i/o$.

If the radiation which reaches the surface is not limited to that with paraxial rays, the reflected rays do not diverge from the same point or converge to the same point. Neighboring rays do form an image but rays with greatly different angles of incidence form images at different places and the net result is an image which is smeared and appears fuzzy.

Problem 8.   Consider the surface and source of problem 1. Suppose the light reaching the surface is limited to rays with $-20^{o}\leq\alpha\leq+20^{0}$.

a. Find the extent of the image along the x axis. To do this, find where the reflection of the two limiting rays ($\alpha=-20^{o}$ and $+20^{o}$) cross that axis and where a paraxial ray ($\alpha=0.5^{o}$) crosses. Use the results of problems 1 and 3.

b. Does the image become more or less fuzzy when the object is nearer the vertex? Put the vertex at $x_{v}=0.02$ m and again use rays with $\alpha=-20^{o}$ and $+20^{o}$. To find the image for paraxial rays, find where the reflection of the incident ray with $\alpha=0.5^{o}$ crosses the axis.

Chapter 20

GEOMETRICAL OPTICS: REFRACTION

Refraction of plane and spherical waves at both plane and spherical boundaries is discussed. Programs to find the directions of refracted rays are developed and used to investigate the passage of light through prisms and lenses. The formation of images and the validity of the thin lens equation for paraxial rays are covered in problems. This material is meant to be used in association with Chapters 43 and 44 of PHYSICS, Chapter 39 of FUNDAMENTALS OF PHYSICS, or similar sections of other texts.

20.1   Refraction at a Plane Boundary

When a wave is incident on a boundary which separates two media, there is usually a wave transmitted into the second medium as well as a reflected wave. The transmitted wave, in general, moves in a direction which is different from that of the incident wave and the wave is said to be refracted. We treat refraction from the point of view of ray optics.

The geometry for a plane boundary is shown in Fig. 20-1. The boundary is perpendicular to the x,y plane and the incident ray lies in that plane. Only the incident and refracted rays are shown, with $\vec{r}_1$ along the incident ray and $\vec{r}_2$ along

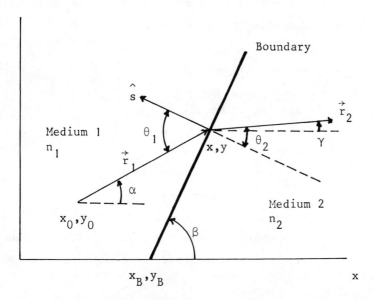

Figure 20-1.   Refraction at a plane boundary.

the refracted ray. The incident ray passes through x_0, y_0, makes the angle α with the x axis, and strikes the boundary at x,y. In the diagram, α increases in the counterclockwise direction. The boundary is described by giving the coordinates x_B, y_B of any point on it and the angle β it makes with the x axis. The unit normal to the boundary is denoted by $\hat{s}$ and the incident ray makes the angle θ_1 with it. The refracted ray makes the angle θ_2 with $\hat{s}$ and the angle γ with the x direction.

The coordinates of the point x,y are found in exactly the same way as was done in section 19.2 and the results are

$$x = x_0 + \ell\cos\alpha \qquad (20\text{-}1)$$

and
$$y = y_0 + \ell\sin\alpha, \qquad (20\text{-}2)$$

with
$$\ell = \frac{(x_B - x_0)\sin\beta + (y_0 - y_B)\cos\beta}{\sin(\beta - \alpha)}. \qquad (20\text{-}3)$$

The direction of the refracted ray is found with the help of the law of refraction or Snell's law:

$$n_1\sin\theta_1 = n_2\sin\theta_2, \qquad (20\text{-}4)$$

where n_1 is the index of refraction for medium 1 (the medium of the incident ray) and n_2 is the index of refraction for medium 2 (the medium of the refracted ray).

If $\hat{r}_2$ is a unit vector in the direction of the refracted ray, then its component in the direction perpendicular to $\hat{s}$ is equal to the same component of $\hat{r}_1$ multiplied by the ratio of the indices of refraction. That is

$$(\hat{r}_2)_{\perp} = (n_1/n_2)(\hat{r}_1)_{\perp}. \qquad (20\text{-}5)$$

This is true since $(\hat{r}_2)_{\perp} = \sin\theta_2$ and $(\hat{r}_1)_{\perp} = \sin\theta_1$. We write

$$\hat{r}_2 = (n_1/n_2)\hat{r}_1 - \vec{\xi} \qquad (20\text{-}6)$$

where $\vec{\xi}$ is a vector to be found. Since the first term produces the correct value for the component of $\hat{r}_2$ in the direction perpendicular to $\hat{s}$, it must be that $\vec{\xi}$ lies along $\hat{s}$ and $\vec{\xi}$ must be chosen so that Eqn. 20-6 produces the correct value for the component of $\hat{r}_2$ along $\hat{s}$. Now the component of $\hat{r}_2$ along $\hat{s}$ is $-\cos\theta_2$ and the component of $\hat{r}_1$ along $\hat{s}$ is

$- \cos\theta_1$ so

$$\hat{r}_2 = (n_1/n_2)\hat{r}_1 + \left[(n_1/n_2)\cos\theta_1 - \cos\theta_2\right]\hat{s}. \qquad (20\text{-}7)$$

The first term in the brackets cancels the component of $(n_1/n_2)\hat{r}_1$ along $\hat{s}$ and the second term adds the component of $\hat{r}_2$ along $\hat{s}$.

$\cos\theta_2$ can be found in terms of $\cos\theta_1$. Use the trigonometric identity $\cos^2A + \sin^2A = 1$ and the law of refraction, as follows:

$$\cos\theta_2 = (1-\sin^2\theta_2)^{\frac{1}{2}} = \left[1 - \frac{n_1^2}{n_2^2}\sin^2\theta_1\right]^{\frac{1}{2}}$$

$$= \left[1 - \frac{n_1^2}{n_2^2}(1 - \cos^2\theta_1)\right]^{\frac{1}{2}}. \qquad (20\text{-}8)$$

Here the positive root is used since $\cos\theta_2$ must be positive. $\cos\theta_1$ is found, as it was in section 19-2, from

$$\cos\theta_1 = -\hat{r}_1 \cdot \hat{s} = \sin(\beta-\alpha). \qquad (20\text{-}9)$$

Since $\hat{r}_1 = \hat{i}\cos\alpha + \hat{j}\sin\alpha$ and $\hat{s} = -\hat{i}\sin\beta + \hat{j}\cos\beta$, the x component of $\hat{r}_2$ is

$$(\hat{r}_2)_x = (n_1/n_2)\cos\alpha - \left[(n_1/n_2)\cos\theta_1 - \cos\theta_2\right]\sin\beta, \qquad (20\text{-}10)$$

its y component is

$$(\hat{r}_2)_y = (n_1/n_2)\sin\alpha + \left[(n_1/n_2)\cos\theta_1 - \cos\theta_2\right]\cos\beta, \qquad (20\text{-}11)$$

and

$$\gamma = \tan^{-1}\frac{(n_1/n_2)\sin\alpha + \left[(n_1/n_2)\cos\theta_1 - \cos\theta_2\right]\cos\beta}{(n_1/n_2)\cos\alpha - \left[(n_1/n_2)\cos\theta_1 - \cos\theta_2\right]\sin\beta}. \qquad (20\text{-}12)$$

Again we take special precautions if the denominator vanishes. Then $\gamma=+90^\circ$ or -90°, depending on the sign of the numerator. If the denominator does not vanish, we use Eqn. 20-12 but make sure the cosine of the result matches the denominator in sign. If it does not, we add 180° to the result.

358

There is one more precaution which we must take. For some situations, the quantity inside the square brackets of Eqn. 20-8 is negative. This occurs if $|(n_1/n_2)\sin\theta_1|>1$. In this event, only reflection occurs and there is no refracted wave. The incident radiation is said to be totally reflected. The phenomenon is called total internal reflection and the smallest angle of incidence for which it occurs is called the critical angle. For total reflection to occur, medium 1 must be optically denser than medium 2 $(n_1>n_2)$.

Fig. 20-2, shown on two pages, gives the flow chart. As for the reflection programs, X1 and X2 contain the x and y coordinates, respectively, of a point on the incident ray and X3 contains the angle α. X4 and X5 contain the x and y coordinates, respectively, of a point on the boundary and X6 contains the angle β between the boundary and the x axis. X10 contains the index of refraction for the medium of the incident ray (to the left of the boundary in Fig. 20-1) and X11 contains the index of refraction for the medium of the refracted ray.

Except for the entry of the indices of refraction, lines 100 through 160 are exactly like the corresponding lines shown in Fig. 19-2. The point of intersection of the incident ray and the boundary are calculated and, at line 160, the coordinates are displayed. x is placed in X1 and y is placed in X2 in preparation for consideration of another boundary. If $\sin(\beta-\alpha)\leq 0$, the ray does not strike the boundary, the machine goes to line 300, and 1×10^{60} is displayed to signal this occurrence. On restarting, the machine goes to line 110 to consider the same ray striking a different boundary.

At line 170, the ratio n_1/n_2 is computed and stored in X10. This combination is used several times and it is convenient to calculate its value just once. Also at line 170, the quantity in the brackets of Eqn. 20-8 is evaluated. If it is negative, total internal reflection occurs. The machine then goes to line 350 where 1×10^{80} is displayed as a signal. On restarting, the machine goes to line 100 to consider a new ray.

At line 190, $\cos\theta_2$ is computed, using Eqn. 20-8. It is stored in X8. Also computed is the combination $(n_1/n_2)\cos\theta_1 - \cos\theta_2$, a quantity which occurs in both Eqns. 20-10 and 20-11. At line 200, the x and y components of the unit vector along the refracted ray are computed and stored in X4 and X5 respectively. These calculations follow Eqns. 20-10 and 20-11.

The last lines of the program are like those of Fig. 19-2. If the x component

Figure 20-2.   Flow chart for a program to trace a ray refracted at a plane boundary.

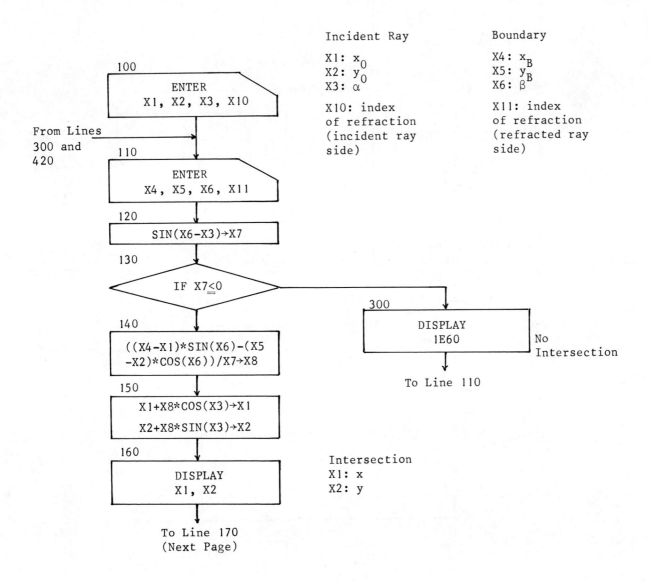

Incident Ray

X1: x_0
X2: y_0
X3: α

X10: index
of refraction
(incident ray
side)

Boundary

X4: x_B
X5: y_B
X6: β

X11: index
of refraction
(refracted ray
side)

100
ENTER
X1, X2, X3, X10

From Lines
300 and
420

110
ENTER
X4, X5, X6, X11

120
SIN(X6-X3)→X7

130
IF X7≤0

300
DISPLAY
1E60

No
Intersection

To Line 110

140
((X4-X1)*SIN(X6)-(X5
-X2)*COS(X6))/X7→X8

150
X1+X8*COS(X3)→X1

X2+X8*SIN(X3)→X2

160
DISPLAY
X1, X2

Intersection
X1: x
X2: y

To Line 170
(Next Page)

360

Fig. 20-2. Cont'd.

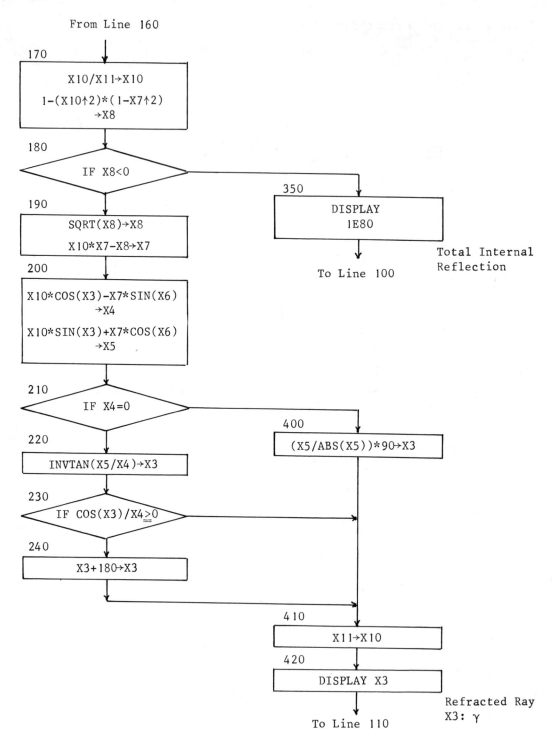

From Line 160

170
X10/X11→X10

1-(X10↑2)*(1-X7↑2)
→X8

180
IF X8<0

190
SQRT(X8)→X8

X10*X7-X8→X7

200
X10*COS(X3)-X7*SIN(X6)
→X4

X10*SIN(X3)+X7*COS(X6)
→X5

210
IF X4=0

220
INVTAN(X5/X4)→X3

230
IF COS(X3)/X4≥0

240
X3+180→X3

350
DISPLAY
1E80

To Line 100

Total Internal
Reflection

400
(X5/ABS(X5))*90→X3

410
X11→X10

420
DISPLAY X3

To Line 110

Refracted Ray
X3: γ

of the unit vector vanishes, γ is set equal to $+90^{\circ}$ or -90°, depending on the sign of the y component. If the x component does not vanish, line 220 produces either γ or $\gamma-180^{\circ}$. The correct choice is made, at lines 230 and 240, by comparing the cosine of the result at line 220 with the x component of the unit vector. In any event, γ is placed in X3 and displayed at line 420.

At line 410, the index of refraction for the medium of the refracted ray is placed in X10. This prepares the machine for the consideration of another boundary, for which this ray is the incident ray. The machine then returns to line 110 where data for the next boundary is entered. The user must remember that, at line 110, the index of refraction for the medium of the new refracted ray must be entered into X11.

<u>Problem 1.</u>   A plane refracting boundary passes through x=0.05 m, y=0 and is oriented at 70° to the x axis. The material to the left has a refractive index of 1 while the material to the right has a refractive index of 1.7. A point source is located at the origin. For each of the ray directions given below, use the program of Fig. 20-2 to find the point of intersection with the boundary and the direction of the refracted ray. Draw a diagram which shows the boundary with its normal, the incident rays, and the refracted rays. For each, use a protractor to measure the angle of incidence and the angle of refraction, then compare $n_1 \sin\theta_1$ with $n_2 \sin\theta_2$.
a. $\alpha = -50^{\circ}$.
b. $\alpha = -20^{\circ}$.
c. $\alpha = 0$.
d. $\alpha = +20^{\circ}$.

Notice that the rays are bent toward the normal to the surface. This occurs when $n_1 < n_2$.

<u>Problem 2.</u>   A plane refracting surface passes through x=0.05 m, y=0 and is oriented at 70° to the x axis. The material to the left has a refractive index of 1.7 while the material to the right has a refractive index of 1. A point source is located at the origin. For each of the ray directions given below, use the program of Fig. 20-2 to find the point of intersection with the boundary and, if refraction occurs, to find the direction of the refracted ray. Draw a diagram which shows the boundary with its normal, the incident rays, and the refracted rays. For each refracted ray, use a protractor to measure the angle of incidence and the angle of refraction, then compare $n_1 \sin\theta_1$ with $n_2 \sin\theta_2$.

362

a. $\alpha = -50^\circ$.

b. $\alpha = -20^\circ$.

c. $\alpha = 0$.

d. $\alpha = +20^\circ$.

Notice that the rays are bent away from the normal. This occurs, quite generally, when $n_1 > n_2$.

Problem 3.   For the situation described in problem 2, find the critical angle for total reflection. Do this by starting with $\alpha = 10^\circ$, then increasing α by 5° until no refraction occurs. This gives a 5° interval which straddles the desired value of α. Try the angle at the midpoint of the interval to see in which half the critical ray lies and use this half as the new interval. Continue to halve the interval until the direction of the critical ray is found to within 0.1°, then calculate the critical angle θ_c. Use $\theta_1 = \alpha - \beta + 90^\circ$. Since $\sin\theta_1 = (n_2/n_1)\sin\theta_2$ and $\sin\theta_2 = 1$ for the critical angle, $\sin\theta_c = n_2/n_1$. Compare your result found using the program with $\sin^{-1}(n_2/n_1)$.

The program of Fig. 20-2 can be used to trace rays which are refracted at more than one surface. After the direction of a refracted ray is found, the machine returns to line 110 to consider the next surface. A point on the ray (the point of intersection with the first surface) and the angle between the ray and the x axis are already stored in the appropriate memories, as is the index of refraction for the medium of the ray.

We first consider a ray refracted by two parallel boundaries of a piece of glass with a rectangular cross section. See Fig. 20-3. Air surrounds the glass so that the wave travels first through air, then through the glass, and finally through air again. If d is the thickness of the glass and the first surface goes through the point x_B, y_B, then the second surface goes through $x_B + d\sin\beta$, $y_B - d\cos\beta$.

The ray emerges from the second surface parallel to the direction with which it strikes the first, but it is not along the same line. Rather, the line of the ray is displaced in a direction perpendicular to the ray. The extent of the displacement depends on the angle of incidence and on the index of refraction of the glass and it is a somewhat complicated function of these quantities. However, if the angle of incidence is small, the displacement is nearly proportional to the angle of incidence.

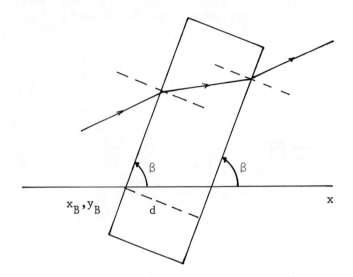

Figure 20-3. A ray passing through a piece of glass with rectangular cross section.

<u>Problem 4.</u>   Suppose a glass is 0.1 m thick, has an index of refraction of 1.65, and is situated so that the first side encountered by the ray passes through $x_B=0$, $y_B=0$.   A point source is at $x=-0.5$ m, $y=0$.   The glass is surrounded by air.

a.   Consider the ray along the x axis ($\alpha=0$) and plot the displacement of the emerging ray from the x axis as a function of the angle β between the surface and the x axis.   The displacement is simply the y coordinate of the intersection of the ray with the second surface.   Start with $\beta=90^O$ and decrease it by 5° each time, until $\beta=45^O$ is reached.   Verify that each ray emerges parallel to its line of incidence and that, for β near 90°, the displacement of the ray is a linear function of β.

b.   Is an image formed by the paraxial rays?   Take $\beta=87^O$ and trace rays with $\alpha=2^O$, 2.5^O, 3°, 3.5^O, and 4°.   Graphically determine if a sharp virtual image is formed by extending the rays backward to see if they intersect at a point.   Why is the image not sharp?   The emerging rays are all parallel to incident rays which diverge from a single point.

The program is now used to study the passage of light through a prism.   The cross section of a prism is triangular in shape, as shown in Fig. 20-4.   It is usual to give the. angle of incidence θ_1 of the ray on the first surface and the prism angle A. If α is chosen to be 0 for the incident ray, then $\beta_1=90^O-\theta_1$ and, regardless of the value of α, $\beta_2=\beta_1+A$.

364

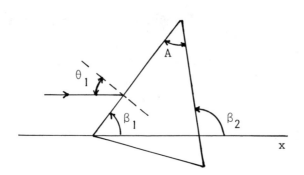

Figure 20-4.   A ray incident
on a prism.

The angle of deviation δ is the angle between the emerging ray and a line
parallel to the incident ray.   In general, it is given by δ=γ-α and, for α=0, by δ=γ.
For purposes of finding the index of refraction of the prism material, the prism is
oriented so that δ is a minimum.   When this occurs,

$$\frac{n_2}{n_1} = \frac{\sin\frac{1}{2}(A+\delta_m)}{\sin\frac{1}{2}A}$$

(20-13)

where δ_m is the angle of minimum deviation, n_2 is the index of refraction of the prism
material, and n_1 is the index of refraction of the surrounding material.   The prism
angle and the angle of minimum deviation are measured, then the values are used in Eqn.
20-13 to calculate n_2/n_1.

Problem 5.   A prism with index of refraction 1.6 and angle A=35° is situated with this
vertex at x=0, y=0.03 m.   A ray is incident in air, along the x axis.
a.   Search for the angle of minimum deviation.   Start with β_1=50° and find the
      direction of the emerging ray.   Increase β_1 by 5° and again find the direction of
      the emerging ray.   Observe that δ decreases in magnitude.   Continue to increment
      β_1 until δ starts to increase.   Then try increments of 1° until you have found the
      minimum to within 1°.
b.   Calculate $\sin\left[\frac{1}{2}(A+\delta_m)\right]/\sin(\frac{1}{2}A)$ and compare with n_2/n_1.
c.   For the condition of minimum deviation, make a diagram showing the prism properly
      oriented, the incident ray, and its path through the prism.   Note that the path is
      symmetric about the line which bisects the apex angle.   In the prism, it is
      perpendicular to this line.   Verify that the angle of refraction at the first

surface is equal to the angle of incidence at the second surface and that they are both equal to A/2. Use data generated by the program. Because of error in δ_m, the verification is not exact.

d. Assume the equality described in part c and analytically derive Eqn. 20-13 starting with the law of refraction.

Problem 6.   A prism with cross section in the shape of a 45° right triangle is oriented with the right angle at the origin and the 45° vertices at x=0, y=0.1 m and x=0.1 m, y=0, respectively. Consider rays incident on the surface which is parallel to the y axis and which then strike the surface forming the hypotenuse. All such rays incident in a downward direction ($\alpha<0$) suffer total internal reflection at the second surface while those which are incident in an upward direction ($\alpha>0$) are refracted at the second surface.

a. What is the index of refraction for this prism? Answer this question analytically.
b. Use the ray tracing program to test your answer. Consider rays which intersect the first surface at x=0, y=0.05 m with $\alpha=-30^o$, -1°, $+1^o$, and $+30^o$. Find the angle of deviation for those rays which are refracted at the second surface.

20.2   Refraction at a Spherical Boundary

The geometry is shown in Fig. 20-5. The vector $\vec{r}_1$ is along the incident ray, $\vec{r}_2$ is along the refracted ray, and these rays make angles of α and γ, respectively, with the x axis. $\hat{s}$ is a unit vector which is normal to the surface at the intersection and which points into the region of the incident ray. For refraction problems, the radius of the boundary is taken to be positive for surfaces which are convex when viewed from the medium of the incident ray and taken to be negative for surfaces which are concave when viewed from that medium. This convention produces a different sign for R than was used for reflection problems but it is consistent with the general rule that surfaces which produce converging rays are called positive and surfaces which produce diverging rays are called negative.

The analysis to find the intersection point follows that of section 19.3. The coordinates of that point are

$$x = x_0 + \ell\cos\alpha \qquad (20-14)$$

and
$$y = y_0 + \ell\sin\alpha, \qquad (20-15)$$

366

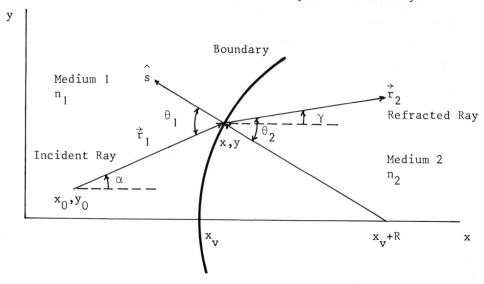

Figure 20-5. Refraction at a
spherical boundary.

where
$$\ell = \left[(x_v + R - x_0)\cos\alpha - y_0\sin\alpha\right] - R\cos\theta_1 \tag{20-16}$$

and
$$\cos\theta_1 = (1/|R|)\left\{\left[(x_v + R - x_0)\cos\alpha - y_0\sin\alpha\right]^2 - (x_v + R - x_0)^2 - y_0^2 + R^2\right\}^{\frac{1}{2}}. \tag{20-17}$$

These equations are the same as analogous equations in section 19.3, except that the sign in front of R is changed wherever it occurs. For the same incident ray, the same radius, and the same type boundary, the two sets of equations produce the same coordinates for the intersection point. The incident ray must travel from left to right, in the diagram, for these equations to be valid.

Just as for refraction at a plane surface, $n_1\sin\theta_1 = n_2\sin\theta_2$, where θ_1 and θ_2 are measured relative to the normal to the surface. The unit vector along the refracted ray is again given by

$$\hat{r}_2 = (n_1/n_2)\hat{r}_1 + \left[(n_1/n_2)\cos\theta_1 - \cos\theta_2\right]\hat{s}, \tag{20-18}$$

an equation which is identical to Eqn. 20-7.

When the substitutions $\hat{r}_1 = \hat{i}\cos\alpha + \hat{j}\sin\alpha$ and $\hat{s} = (1/R)\left[(x - x_v - R)\hat{i} + y\hat{j}\right]$ are made in Eqn. 20-18, the x component of $\hat{r}_2$ is found to be

$$(\hat{r}_2)_x = (n_1/n_2)\cos\alpha + \left[(n_1/n_2)\cos\theta_1 - \cos\theta_2\right](x-x_v-R)/R \qquad (20\text{-}19)$$

and the y component is found to be

$$(\hat{r}_2)_y = (n_1/n_2)\sin\alpha + \left[(n_1/n_2)\cos\theta_1 - \cos\theta_2\right]y/R. \qquad (20\text{-}20)$$

Here $\cos\theta_2$ is calculated using

$$\cos\theta_2 = \left[1 - \frac{n_1^2}{n_2^2}(1 - \cos^2\theta_1)\right]^{\frac{1}{2}}. \qquad (20\text{-}21)$$

The flow chart for the program is shown in Fig. 20-6. The logic and sequence of steps are like those of the program shown in Fig. 19-5. The x and y coordinates of a point on the incident ray, the angle this ray makes with the x axis, and the index of refraction of the material to the left of the surface are entered at line 100, into X1, X2, X3, and X10, respectively. The coordinate of the surface vertex, the radius of the surface, and the index of refraction of the material to the right of the surface are entered, at line 110, into X4, X5, and X11, respectively. The radius R is positive for convex surfaces and negative for concave surfaces.

At line 120, the combination x_v+R-x_0 is calculated and stored in X6, the combination $(x_v+R-x_0)\cos\alpha - y_0\sin\alpha$ is calculated and stored in X7, and $R^2\cos^2\theta_1$ is calculated and stored in X8. Eqn. 21-17 is used. If this last quantity is negative, the ray does not intersect the surface and the machine proceeds to line 300 where the signal 1×10^{60} is displayed.

At line 140, $\cos\theta_1$ and ℓ are calculated and stored in X8 and X9 respectively. The coordinates of the intersection point are computed at line 150 and stored in X1 and X2. They are then displayed.

The ratio n_1/n_2 of the indices of refraction is computed at line 170 and stored in X10. Also computed at that line is $\cos^2\theta_2$; it is placed in X9. If this quantity is negative, total internal reflection occurs and the machine proceeds to line 350, where 1×10^{80} is displayed as a signal.

At line 190, $\cos\theta_2$ and $(n_1/n_2)\cos\theta_1 - \cos\theta_2$ are computed and stored in X9 and X8 respectively. The x and y components of $\hat{r}_2$ are calculated at line 200 and stored in X7 and X8 respectively. The rest of the program, which is concerned with the

368

Figure 20-6.   Flow chart for a program to trace a ray refracted at a
spherical boundary.

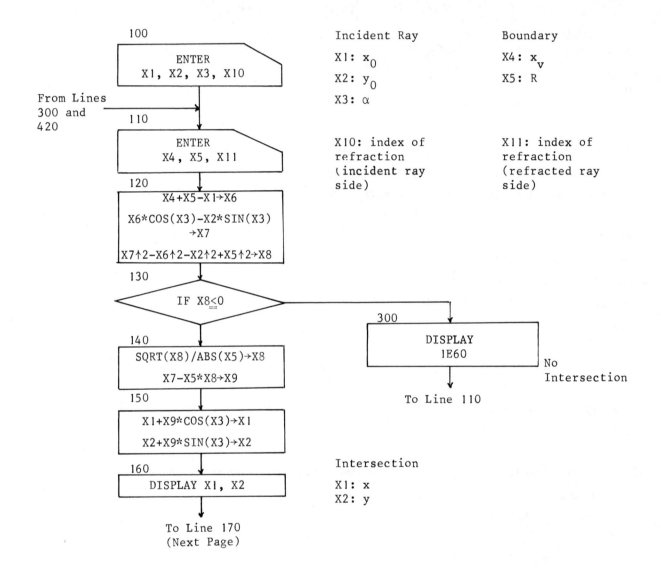

369

Fig. 20-6.   Cont'd.

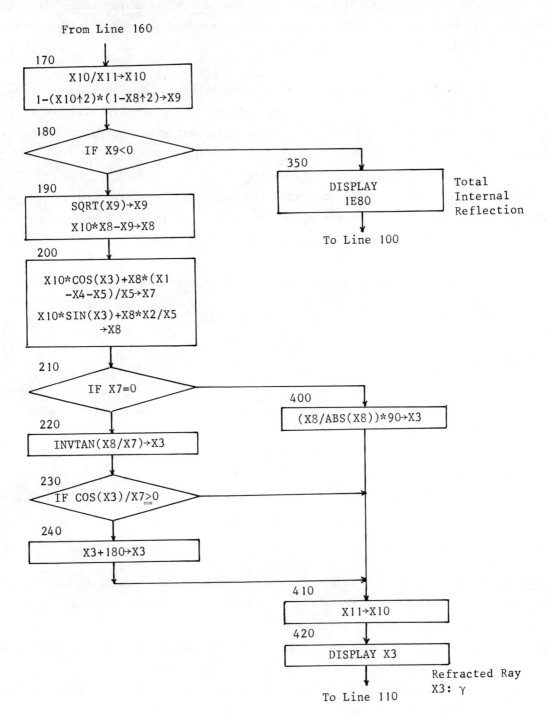

calculation of γ, is like the similar section of the program of Fig. 20-2. After displaying γ, in X3, the machine returns to line 110 where information about another boundary can be entered. Information about the ray and the index of refraction have already been placed in appropriate storage locations.

More than one boundary may actually be present. An additional boundary may also be inserted into the problem for the sole purpose of determining the coordinates of a point on the last refracted ray. These coordinates are useful when the ray is traced.

The intersection x_c of the ray with the x axis can be found using $x_c = x - y/\tan\gamma$ where x and y are the coordinates of the intersection with the last surface. For more details refer to the discussion just prior to problem 1 of section 19.3.

The following two problems can be used to test the program.

Problem 1. A convex surface has radius R=0.18 m and its vertex is at the origin. The material to the left of the surface is air (n_1=1) and the material to the right has an index of refraction of 1.45.

a. Use the program of Fig. 20-6 to trace the following rays, all of which go through x=-0.3 m, y=0: α=-20°, -10°, 0, +10°, and +20°.

b. On a graph, draw the surface to scale and plot the rays and the normals at the points of intersection.

c. With a protractor, measure the angles of incidence and refraction for each ray. Verify, in each case, that $n_1\sin\theta_1 = n_2\sin\theta_2$.

Problem 2. A concave surface has radius R=0.18 m and its vertex is at the origin. The material to the left of the surface is air (n_1=1) and the material to the right has index of refraction 1.45.

a. Use the program of Fig. 20-6 to trace the following rays, all of which pass through x=-0.3 m, y=0: α=-20°, -10°, 0, +10°, and +20°.

b. On a graph, draw the surface to scale and plot the rays and the normals to the surface at the points of intersection.

c. With a protractor, measure the angles of incidence and refraction for each ray. Verify, in each case, that $n_1\sin\theta_1 = n_2\sin\theta_2$.

Spherical refracting surfaces cause rays, initially diverging from a source, to form images. In some cases, the refracted rays converge to a real image while, in other cases, the refracted rays diverge from a virtual image. Since all refracted rays are to the right of the boundary, real images are formed to the right and virtual images are formed to the left when the rays are incident from the left. If the incident rays, from a point source, are paraxial, the refracted rays converge to or diverge from nearly the same point and the image is sharp. Non-paraxial rays form fuzzy images when refracted at spherical surfaces.

If the surface is convex and $n_2 > n_1$, then a real image is formed for an object on the axis far from the vertex and a virtual image is formed for an object on the axis close to the vertex. There is a point on the axis such that, if a point source is placed there, all paraxial rays become, after refraction, parallel to the axis. This point is called the object focus or front focus of the surface and the image is said to be at infinity. As a point source is moved along the axis from far away toward the vertex, the image is at first real and fairly close to the vertex. Then the refracted rays become more nearly parallel to the axis and the image recedes. When the source is at the object focus, the image undergoes the change from real to virtual. When the source is just inside the object focus, the image is virtual and far from the surface, on the same side as the source. As the source moves still closer to the vertex, the image also moves in toward the vertex.

For a given source position, the position of the image formed by paraxial rays can be found geometrically. Object and image distances are related by

$$\frac{n_1}{o} + \frac{n_2}{i} = \frac{n_2 - n_1}{R} \ . \tag{20-22}$$

Here o is the distance from the source to the vertex and, by convention, it is positive for real sources, from which rays diverge. i is the distance from the image to the vertex and it is positive for real images, toward which the refracted rays converge, and negative for virtual images, from which the refracted rays diverge.

If $i = \infty$, $o = f_o$, the object focal length. So $n_1/f_o = (n_2 - n_1)/R$ or

$$f_o = \frac{n_1}{n_2 - n_1} R. \tag{20-23}$$

When the object is far away, the point toward which the refracted rays converge is called the image focus or back focus and its distance from the vertex is denoted by f_i. According to Eqn. 20-22,

$$f_i = \frac{n_2}{n_2 - n_1} R. \tag{20-24}$$

Incident rays parallel to the optic axis converge to this point after refraction. Note that, in general, the object and image focal points are different distances from the vertex. They are always on opposite sides of the vertex.

Problem 3.   A convex surface has a radius of 0.023 m and vertex at the origin. Take $n_1=1$, $n_2=1.4$, and, for each of the following source positions on the x axis, use the program of Fig. 20-6 to find where paraxial rays cross the x axis. This point is the position of the image. To insure that paraxial rays are used, take $\alpha=\arc\tan(1\times10^{-4}/|x_0|)$ or 1×10^{-2} degrees, whichever is less.

a. $x_0=-200$ m, $y_0=0$. The source is far away from the vertex and the image should be near the image focus. Compare your answer with that found using Eqn. 20-24. Is the image real or virtual?

b. $x_0=-0.5$ m, $y_0=0$. What has happened to the image position? Compare the image position produced by the program with that found using Eqn. 20-22. Is the image real or virtual?

c.   Use Eqn. 20-23 to compute f_o. Place the source at the object focus and use the program to verify that the refracted rays are parallel to the x axis. Expect some round off error.

d. $x_0=-0.002$ m, $y_0=0$. What has happened to the image position? Compare the image position with that found using Eqn. 20-22. Is the image real or virtual?

What happens if the surface is concave and R is negative? Then both f_o and f_i are negative. A negative image focal length is not hard to understand. Paraxial rays with $\alpha=0$ diverge after refraction at the concave surface and f_i gives the position of the point from which they diverge. It is on the same side of the surface as the incident rays.

The object focal length is also negative. It is to the right of the surface and represents the position of a virtual object, one toward which the incident rays converge. Incident rays, converging toward the object focal point are refracted so that they become parallel to the optic axis.

Problem 4. A concave surface has a radius of 0.023 m and vertex at the origin. Take $n_1=1$, $n_2=1.4$, and, for each of the following source positions on the x axis, use the program of Fig. 20-6 to find where paraxial rays cross the x axis. This is the position of the image. Take α to be the smaller of arc $\tan(1\times10^{-4}/|x_0|)$ and 1×10^{-2} degrees in all cases except part d.

a. $x_0=-200$ m, $y_0=0$. This ray should be refracted radially outward from the image focus. Compare your answer with $f_i=Rn_2/(n_2-n_1)$. Is the image real or virtual?

b. $x_0=-0.5$ m, $y_0=0$. What has happened to the image position? Use Eqn. 20-22 to find the image position and compare with the answer produced by the machine. Is the image real or virtual?

c. $x_0=-0.002$ m, $y_0=0$. What has happened to the image position? Use Eqn. 20-22 to find the image position and compare with the answer produced by the machine. Is the image real or virtual?

d. Use Eqn. 20-23 to compute f_o. Place a point source at $x_0=-0.1$ m, $y_0=1\times10^{-4}$ m and choose α so that the incident ray, if extended, passes through the object focus. The object focus is to the right of the surface. Trace the ray and find γ.

The image of an extended object can be found by locating the images of various points on the object. We consider an object in the shape of an arrow with the tail on the optic axis and extending in the positive y direction, perpendicular to the optic axis. Images of the head and tail are found and these images are connected by a straight line to form the image of the arrow. It is of interest to find out if the image is erect or inverted with respect to the object and to find the magnification for the surface. The magnification is the ratio of the image length to the object length. It is considered positive if the image is erect and negative if the image is inverted. For paraxial rays, the magnification is given by the ratio -i/o.

Problem 5.   A 0.001 m arrow is placed perpendicular to the x axis, in front of a convex refracting surface with radius 0.023 m and with vertex at the origin. Take $n_1=1$ and $n_2=1.4$. For each of the following positions of the object, trace two paraxial rays from the head and find their intersection after they are refracted. This is the image of the head. The image of the tail was found in problem 3. In each case, tell if the image is real or virtual, tell if the image is erect or inverted, and calculate the magnification.

a. $x_0=-200$ m. Use rays with $\alpha=0$ and -5×10^{-4} degrees, respectively.

b. $x_0=-0.5$ m. Use rays with $\alpha=0$ and -1×10^{-2} degrees. respectively.

c. $x_0=-0.002$ m. Use rays with $\alpha=0$ and -1×10^{-2} degrees, respectively.

If the rays are not paraxial, the image is smeared. Some indication of the smearing can be obtained by working the following problem.

Problem 6.   Consider the convex surface of problem 3 ($x_v=0$, R=0.023 m, $n_1=1$, $n_2=1.4$). A point source is at $x_0=-0.15$ m, $y_0=0$. Trace the rays with $\alpha=6^{\circ}$, 4°, 2°, -2°, -4°, and -6°. On a graph, plot the rays and place circles at points of intersection. The positions of the circles give an indication of the smearing of the image when a 12° cone of light is incident on the surface.

20.3   Lenses

A lens is a piece of refracting material shaped to focus rays of light. Generally, it consists of two surfaces and is designed to achieve a sharp image for sources at a wide variety of positions, to accept a large amount of light, and to produce an undistorted image of an extended source. Not all of these goals can be met simultaneously and which predominates in the design depends on the application for which the lens is intended. One of the most important practical applications of geometrical optics is the design of lenses.

Many modern lenses have surfaces which are not spherical in shape. Attempts to achieve the goals mentioned above, and others, have led to the use of non-spherical surfaces and to systems of lenses, in which one lens compensates for deficiencies in another. In this introduction, we concentrate on lenses with spherical surfaces.

The program of Fig. 20-6 can be used to trace rays from source to image, through the lens. Information about the second surface is entered once the inter- section of the ray with the first surface and the direction of the ray in the lens have been found. For all our applications, the index of refraction for the lens material is given and this is entered in X11 before the first surface is considered. The lens is in air and an index of refraction of 1 is entered into X10 for the first surface and into X11 for the second surface. The index of refraction for the lens is automatically transferred from X11 to X10 after the calculation for the first surface has been carried out.

A warning is in order about the sign convention for the surface radii. A surface is convex, with R positive, or concave, with R negative, according to its curvature as seen from the point of view of the incoming ray. For example, so-called double convex lenses have both sides bowed outward but the first surface has a positive radius while the second surface has a negative radius. The second surface is concave to the ray which strikes it. Similarly, a double concave lens has both sides bowed inward but the first surface has a negative radius while the second has a positive radius.

We begin by investigating the influence of various types of lenses on paraxial rays. The notation used in the problems is: the x axis is again taken to be the optic axis, the first surface has its vertex at x_{v1} and has radius R_1, the second surface has its vertex at x_{v2} and has radius R_2.

For many problems, a point source is placed on the optic axis. The image formed by paraxial rays can be found by tracing a single paraxial ray and finding its intersection with the optic axis. When the source is off the axis, of course, two paraxial rays must be traced and their intersection with each other found. The program of Fig. 19-2 can be used to do this. One ray plays the role of a plane mirror and the intersection point with the other ray is found. In some cases, one or both of the rays must be reversed by adding 180° to γ. This is because, as the program is written, only one side of the plane mirror reflects.

Problem 1.   A point source is at $x_0=-2$ m, $y_0=0$, in front of a double convex lens with index of refraction 1.47. The vertex of the first surface is at $x_{v1}=0$ and this surface has a radius of 0.2 m ($R_1=+0.2$ m) while the vertex of the second surface is at $x_{v2}=0.008$ m and this surface has a radius of 0.6 m ($R_2=-0.6$ m).

a. Consider two paraxial rays, with $\alpha=0.005°$ and $0.001°$, respectively. Find the intersections of the refracted rays with the x axis and check to see that they cross at very nearly the same place. Is the image real or virtual?

b. Repeat the calculation of part a, but with the source at $x_0=-0.5$ m, $y_0=0$. In what direction has the image moved? Is it real or virtual?

c. Repeat the calculation of part a, but with the source at $x_0=-0.1$ m, $y_0=0$. In what direction has the image moved? Is it real or virtual?

Notice that there are object distances for which the image is real and object distances for which the image is virtual. The image distance for paraxial rays can be found analytically, but we shall not pursue the matter here.

Problem 2. A point source is at $x_0=-2$ m, $y_0=0$, in front of a convex-concave lens with index of refraction 1.47. The vertex of the first surface is at $x_{v1}=0$ and this surface has a radius of 0.2 m ($R_1=+0.2$ m) while the vertex of the second surface is at $x_{v2}=0.008$ m and this surface has a radius of 0.6 m ($R_2=+0.6$ m). The lens is the same as that of problem 1, except that the second surface is turned around.

a. Consider the paraxial ray with $\alpha=0.001°$ and find the image of the source. Is it real or virtual?

b. Repeat the calculation of part a, but with the source at $x_0=-0.5$ m, $y_0=0$. In what direction has the image moved? Is it real or virtual?

c. Repeat the calculation of part a, but with the source at $x_0=-0.1$ m, $y_0=0$. In what direction has the image moved? Is it real or virtual?

Problem 3. A point source is at $x_0=-2$ m, $y_0=0$, in front of a double concave lens with index of refraction 1.47. The vertex of the first surface is at $x_{v1}=0$ and this surface has a radius of 0.2 m ($R_1=-0.2$ m) while the vertex of the second surface is at $x_{v2}=0.008$ m and this surface has a radius of 0.6 m ($R_2=+0.6$ m). The lens is the same as that of problem 2, except that the first surface is turned around.

a. Consider the paraxial ray with $\alpha=0.001^{\circ}$ and find the image of the source. Is it real or virtual?

b. Repeat the calculation of part a, but with the source at $x_0=-0.5$ m, $y_0=0$. In what direction has the image moved? Is it real or virtual?

c. Repeat the calculation of part a, but with the source at $x_0=-0.1$ m, $y_0=0$. In what direction has the image moved? Is it real or virtual?

A lens has two focal points: an object focus, such that rays from a point source placed there are parallel to the optic axis after refraction and an image focus to which rays converge if they are incident parallel to the optic axis.

Problem 4.

a. Find the focal points for the lens of problem 1. Find the object focus by trial and error: move the source along the x axis until emerging rays are nearly parallel to that axis. To obtain three figure accuracy, find x_0 so that a change of 1 in the third significant figure changes the sign of γ. Find the image focus by considering an incident ray with $\alpha=0$ and finding where the emerging ray crosses the x axis. In each case, use paraxial rays, never more than 1×10^{-5} m from the x axis.

b. Suppose the lens were turned around so that the 0.6 m radius surface is the first surface struck by the incident rays and suppose a point source were placed a distance f_i, as found in part a, in front of the lens. What angles do the emerging rays make with the x axis after refraction? Make a guess, based on your physical intuition, then check the guess by making a calculation.

If the rays which are not paraxial are allowed to strike the lens and contribute to the formation of the image, the image is smeared out.

Problem 5.   Consider the lens of problem 1 and, for each of the object positions given below, trace the ray with $\alpha=3^{\circ}$ and find where it crosses the x axis after

378

passing through the lens. In each case, find the displacement of the point of intersection from the image formed by paraxial rays.

a. $x_0 = -2$ m, $y_0 = 0$.

b. $x_0 = -0.5$ m, $y_0 = 0$.

c. $x_0 = -0.1$ m, $y_0 = 0$.

Lenses are sometimes used to focus rays onto a photographic film, for example. In the following problems, the lens-to-film distance is fixed and, for various object positions on the axis, we find where rays through the outer part of the lens strike the film. The closer this point is to the axis, the better the focus.

Problem 6. Consider the lens of problem 1 and suppose a photographic plate, perpendicular to the x axis, is positioned at the place where paraxial rays from a point source at $x_0 = -0.5$ m, $y_0 = 0$ are focused. See the solution to problem 1 to find where that place is. Consider a ray from the point source which strikes the front surface of the lens at y=0.005 m and, for each of the following source positions, find where the ray strikes the plate after refraction. For these rays, $\alpha = $ arc $\tan(0.005/|x_0|)$.

a. $x_0 = -0.5$ m, $y_0 = 0$.

b. $x_0 = -2$ m, $y_0 = 0$.

c. $x_0 = -4$ m, $y_0 = 0$.

If the aperture in front of the lens is closed down so that the rays received are more nearly paraxial, the image sharpens. Narrowing the aperture increases the depth of field, the range of object distances for which sharp images are produced.

Problem 7. Consider the lens and object distances of problem 6 and find the points of intersection with the photographic plate for rays which strike the lens at y=0.002 m. In each case, calculate the ratio of the image width to the image width for a 0.005 m radius aperture.

The process of narrowing the aperture to improve the image cannot be continued indefinitely. At small apertures, so-called diffraction phenomena predominate and the light tends to spread. These phenomena will be discussed in a later chapter.

If the thickness of the lens is small compared to other relevant distances, such as the focal lengths, the object distance, and the image distance, then it can be ignored for many applications. The lens is then called a thin lens. In that case, the object and image focal lengths have the same value and the object and image distances, for paraxial rays, are related by

$$\frac{1}{o} + \frac{1}{i} = \frac{1}{f}$$ (20-25)

where f is the focal length. Object and image distances are measured from the position of the lens, with o positive for real objects, in front of the lens, and i positive for real images, behind the lens. Again, i is negative for virtual images.

Furthermore, the focal length is related to the radii of curvature and the index of refraction of the lens material by

$$\frac{1}{f} = (n-1)(\frac{1}{R_1} - \frac{1}{R_2})$$ (20-26)

where R_1 is the radius of the first surface and R_2 is the radius of the second surface. The usual sign convention, as given earlier, applies. The lens is assumed to be in air, with index of refraction 1.

Problem 8.   Is the lens of problem 1 a thin lens for the source positions considered in that problem?
a.   Use Eqn. 20-26 to calculate the focal length in the thin lens approximation and compare the result with the two focal lengths found in reponse to problem 4.
b.   Use Eqn. 20-25 to calculate the image positions for paraxial rays from point sources at x_0=-2 m, x_0=-0.5 m, and x_0=-0.1 m, all on the x axis.

Compare the results with the image positions found in answer to problem 1. In each case, calculate the displacement of the image from the position predicted by the thin lens equations.

c. Now consider a lens with the same surfaces but with 0.001 m between vertices. Use the program of Fig. 20-6 to find the image positions for point sources at $x_0 = -2$ m, -0.5 m, and -0.1 m, all or the x axis. Use paraxial rays.

d. Compare the results of part c with the thin lens results of part b. In each case, calculate the displacement of the image from the position predicted by the thin lens equations. Has the narrowing of the lens improved agreement? Has agreement improved more for some object distances than for others?

Chapter 21

MATRICES, WITH APPLICATIONS TO LENS SYSTEMS

Matrix methods for tracing paraxial rays through systems of lenses are described and applied to telescopes, microscopes, and camera lenses. Although the examples given are comparatively simple, these techniques are widely used for the analysis of complicated lens systems. In section 21.1, matrix algebra is discussed and a program for finding the inverse of a matrix and for finding the product of two matrices is given. The problems in section 21.3 may be used in association with laboratory experiments. They may also form the basis for an independent study of lens systems.

21.1   Matrices

A matrix is an array of numbers, usually written in the form of n rows, each containing m entries. Such an array is said to be an n×m matrix. An example of a 3×2 matrix is

$$\begin{pmatrix} 5.1 & 3.2 \\ 7.5 & 1.6 \\ 0 & 3.1 \end{pmatrix}.$$

To be a matrix, an array must also obey certain rules of arithmetic, to be given later.

We deal exclusely with 3 types of matrices: square (the number of columns is the same as the number of rows), column (a single column), and row (a single row). In this chapter, all matrices are 2×2, 2×1, or 1×2. The individual numbers in a matrix are called the elements of the matrix.

For purposes of defining various matrix quantities and mathematical manipulations, standard matrix notation is used. Later, the notation is changed slightly for ease in programming.

A matrix is denoted by a single capital letter, underlined twice: $\underline{\underline{A}}$, for example. An element of $\underline{\underline{A}}$ is denoted by a_{ij}, with the first subscript giving the row and the second subscript giving the column. Thus

382

$$A = \begin{pmatrix} a_{11} & a_{12} \\ a_{21} & a_{22} \end{pmatrix} \tag{21-1}$$

is a 2×2 matrix. Usually only one subscript is given for an element of a column or row matrix. As examples,

$$B = \begin{pmatrix} b_1 \\ b_2 \end{pmatrix} \tag{21-2}$$

and
$$C = (c_1 \quad c_2) \tag{21-3}$$

are column and row matrices, respectively. Both subscripts are given when there might be confusion as to whether the matrix is column or row in form.

Two matrices are equal if each element of one equals the corresponding element of the other. That is, $A=B$ if $a_{ij}=b_{ij}$ for all values of i and j. A and B, of course, must have the same number of rows and the same number of columns.

To be matrices, arrays must obey certain rules of addition, subtraction, and multiplication. Addition is defined for two matrices with the same number of columns and the same number of rows. The sum is another matrix, and any element of the sum is the sum of the corresponding elements of the addends. That is, if $C=A+B$, then

$$c_{ij} = a_{ij} + b_{ij} \tag{21-4}$$

or, written out in full for 2×2 matrices,

$$C = \begin{pmatrix} a_{11}+b_{11} & a_{12}+b_{12} \\ a_{21}+b_{21} & a_{22}+b_{22} \end{pmatrix}. \tag{21-5}$$

Subtraction is defined in an analogous way. The difference between two matrices is the matrix formed by the difference in corresponding elements. If $C=A-B$, then

$$c_{ij} = a_{ij} - b_{ij}, \tag{21-6}$$

or

$$\underline{\underline{C}} = \begin{pmatrix} a_{11}-b_{11} & a_{12}-b_{12} \\ a_{21}-b_{21} & a_{22}-b_{22} \end{pmatrix}.$$ (21-7)

The examples given are for 2×2 matrices, but the definitions are equally valid for any size matrices, including column and row matrices.

Matrix multiplication is a little more complicated. If $\underline{\underline{C}}=\underline{\underline{B}}\cdot\underline{\underline{A}}$, the ij element of $\underline{\underline{C}}$ is formed by considering row i of $\underline{\underline{B}}$ and column j of $\underline{\underline{A}}$. The first element of the row of $\underline{\underline{B}}$ is multiplied by the first element of the column of $\underline{\underline{A}}$, the second element of the row of $\underline{\underline{B}}$ is multiplied by the second element of the column of $\underline{\underline{A}}$. This process is continued until all elements of the row and column have been used. Then the results are summed. For two 2×2 matrices, for example, $c_{12} = b_{11}a_{12} + b_{12}a_{22}$. Notice that the elements of $\underline{\underline{B}}$ are both from the first row, the elements of $\underline{\underline{A}}$ are both from the second column, and, for each term, the column number for the element from $\underline{\underline{B}}$ is the same as the row number for the element from $\underline{\underline{A}}$. The complete matrix is

$$\underline{\underline{C}} = \begin{pmatrix} b_{11} & b_{12} \\ b_{21} & b_{22} \end{pmatrix} \cdot \begin{pmatrix} a_{11} & a_{12} \\ a_{21} & a_{22} \end{pmatrix} = \begin{pmatrix} b_{11}a_{11}+b_{12}a_{21} & b_{11}a_{12}+b_{12}a_{22} \\ b_{21}a_{11}+b_{22}a_{21} & b_{21}a_{12}+b_{22}a_{22} \end{pmatrix}.$$

(21-8)

As an aid in carrying out matrix multiplication, start with the first row of $\underline{\underline{B}}$ and the first column of $\underline{\underline{A}}$. Move the index finger of the left hand from entry to entry along the row and the index finger of the right hand down the column. Move them in unison from element to element and, at each element, find the product of the elements to which the fingers point. When finished, add up the products. This gives c_{11}. To find c_{12}, repeat the process using the first row of $\underline{\underline{B}}$ and the second column of $\underline{\underline{A}}$. c_{21} is found using the second row of $\underline{\underline{B}}$ and the first column of $\underline{\underline{A}}$ while c_{22} is found using the second row of $\underline{\underline{B}}$ and the second column of $\underline{\underline{A}}$.

To form the product, $\underline{\underline{B}}$ must have the same number of columns as $\underline{\underline{A}}$ has rows. It is possible to multiply a square matrix and a column or row matrix:

$$
\begin{pmatrix} a_{11} & a_{12} \\ a_{21} & a_{22} \end{pmatrix} \cdot \begin{pmatrix} b_1 \\ b_2 \end{pmatrix} = \begin{pmatrix} a_{11}b_1 + a_{12}b_2 \\ a_{21}b_1 + a_{22}b_2 \end{pmatrix} \tag{21-9}
$$

and
$$
(c_1 \quad c_2) \cdot \begin{pmatrix} a_{11} & a_{12} \\ a_{21} & a_{22} \end{pmatrix} = (c_1 a_{11} + c_2 a_{21} \quad c_1 a_{12} + c_2 a_{22}). \tag{21-10}
$$

The order of the factors counts. $\underline{A} \cdot \underline{B}$ does not yield the same matrix as $\underline{B} \cdot \underline{A}$. In fact, $\underline{B} \cdot \underline{A}$ is not defined if $\underline{B}$ is a column matrix and $\underline{A}$ is square. $\underline{A} \cdot \underline{B}$ is not defined if $\underline{B}$ is a row matrix and $\underline{A}$ is square. If $\underline{B}$ is a 2×1 column matrix and $\underline{C}$ is a 1×2 row matrix, then $\underline{C} \cdot \underline{B} = c_1 b_1 + c_2 b_2$, a single number, but $\underline{B} \cdot \underline{C}$ is not defined.

A set of linear homogeneous algebraic equations can be written as an equality between a column matrix and the product of a square matrix with another column matrix. For example, consider the set

$$
5.3x_1 + 3.2x_2 = 7.1
$$
$$
1.6x_1 - 6.4x_2 = 5.8 \tag{21-11}
$$

and define the matrices

$$
\underline{A} = \begin{pmatrix} 5.3 & 3.2 \\ 1.6 & -6.4 \end{pmatrix} ,
$$

$$
\underline{X} = \begin{pmatrix} x_1 \\ x_2 \end{pmatrix} ,
$$

and
$$
\underline{C} = \begin{pmatrix} 7.1 \\ 5.8 \end{pmatrix} . \tag{21-12}
$$

Then the set of simultaneous equations can be written

$$\underline{\underline{A}} \cdot \underline{X} = \underline{C}. \tag{21-13}$$

Perform the matrix multiplication $\underline{\underline{A}} \cdot \underline{X}$ according to the rules indicated by Eqn. 21-9. Equate the upper element in the product to the upper element in $\underline{C}$, the lower element in the product to the lower element in $\underline{C}$, and check to see that the original equations are recovered.

Matrix multiplication is associative. That is

$$\underline{\underline{C}} \cdot (\underline{\underline{B}} \cdot \underline{\underline{A}}) = (\underline{\underline{C}} \cdot \underline{\underline{B}}) \cdot \underline{\underline{A}}. \tag{21-14}$$

When three matrices are multiplied, it does not matter whether the third is multiplied by the product of the second and first or the product of the third and second is multiplied by the first.

There is an inverse matrix associated with certain of the square matrices we consider. The matrix which is inverse to $\underline{\underline{A}}$ is denoted by $\underline{\underline{A}}^{-1}$ and it has the property that $\underline{\underline{A}} \cdot \underline{\underline{A}}^{-1}$ and $\underline{\underline{A}}^{-1} \cdot \underline{\underline{A}}$ are the same and each is a matrix with a 1 in every location along the diagonal from the upper left corner to the lower right corner. All other elements are zero. For a 2×2 matrix $\underline{\underline{A}}$,

$$\underline{\underline{A}} \cdot \underline{\underline{A}}^{-1} = \underline{\underline{A}}^{-1} \cdot \underline{\underline{A}} = \begin{pmatrix} 1 & 0 \\ 0 & 1 \end{pmatrix}. \tag{21-15}$$

The matrix on the right side of Eqn. 21-15 is called the unit matrix. In matrix multiplication, it acts like the number 1 does in the multiplication of ordinary numbers. When it is multiplied by any 2×2 matrix $\underline{\underline{A}}$, the result is $\underline{\underline{A}}$.

The inverse is useful, for example, in the solution of a set of simultaneous equations. If $\underline{\underline{A}} \cdot \underline{X} = \underline{B}$, where $\underline{X}$ and $\underline{B}$ are column matrices with $\underline{X}$ unknown and $\underline{B}$ known, $\underline{X}$ can be found by multiplying from the left by $\underline{\underline{A}}^{-1}$:

$$\underline{\underline{A}}^{-1} \cdot (\underline{\underline{A}} \cdot \underline{X}) = \underline{\underline{A}}^{-1} \cdot \underline{B}. \tag{21-16}$$

Since multiplication is associative, $\underline{\underline{A}}^{-1} \cdot (\underline{\underline{A}} \cdot \underline{X}) = (\underline{\underline{A}}^{-1} \cdot \underline{\underline{A}}) \cdot \underline{X} = \underline{X}$, so

$$\underline{X} = \underline{\underline{A}}^{-1} \cdot \underline{B}. \tag{21-17}$$

It is easy to solve for the inverse of a 2×2 matrix in terms of elements of the original matrix. If

$$\underline{\underline{A}} = \begin{pmatrix} a_{11} & a_{12} \\ a_{21} & a_{22} \end{pmatrix} \tag{21-18}$$

and

$$\underline{\underline{A}}^{-1} = \begin{pmatrix} b_{11} & b_{12} \\ b_{21} & b_{22} \end{pmatrix}, \tag{21-19}$$

then

$$\underline{\underline{A}} \cdot \underline{\underline{A}}^{-1} = \begin{pmatrix} a_{11}b_{11}+a_{12}b_{21} & a_{11}b_{12}+a_{12}b_{22} \\ a_{21}b_{11}+a_{22}b_{21} & a_{21}b_{12}+a_{22}b_{22} \end{pmatrix} \tag{21-20}$$

and we wish this to be the unit matrix. So

$$a_{11}b_{11}+a_{12}b_{21} = 1,$$
$$a_{11}b_{12}+a_{12}b_{22} = 0,$$
$$a_{21}b_{11}+a_{22}b_{21} = 0,$$

and

$$a_{21}b_{12}+a_{22}b_{22} = 1. \tag{21-21}$$

These are four simultaneous equations to be solved for the unknowns b_{11}, b_{12}, b_{21}, and b_{22}. The results are

$$b_{11} = a_{22}/(a_{11}a_{22}-a_{12}a_{21}),$$
$$b_{12} = -a_{12}/(a_{11}a_{22}-a_{12}a_{21}),$$
$$b_{21} = -a_{21}/(a_{11}a_{22}-a_{12}a_{21}),$$

and

$$b_{22} = a_{11}/(a_{11}a_{22}-a_{12}a_{21}). \tag{21-22}$$

In each case, the denominator is the determinant of the array $\underline{\underline{A}}$. For $\underline{\underline{A}}$ to have an inverse, its determinant must not vanish.

387

For the applications, we present a program to perform matrix multiplication and to find the inverse of a square matrix. Only 2×2 matrices and 2 element column matrices are considered, since these are the matrices we shall use to study optical systems.

For ease in programming, the notation is changed so that only a single subscript is used. A 2×2 matrix is written

$$A = \begin{pmatrix} a_1 & a_2 \\ a_3 & a_4 \end{pmatrix}. \qquad (21\text{-}23)$$

With the new notation, Eqn. 21-8 for the multiplication of two 2×2 matrices is equivalent to the set of equations

$$c_1 = b_1 a_1 + b_2 a_3,$$
$$c_2 = b_1 a_2 + b_2 a_4,$$
$$c_3 = b_3 a_1 + b_4 a_3,$$

and
$$c_4 = b_3 a_2 + b_4 a_4. \qquad (21\text{-}24)$$

Eqn. 21-9, for the multiplication of a column matrix by a square matrix, is equivalent to the set of equations

$$c_1 = a_1 b_1 + a_2 b_2$$

and
$$c_2 = a_3 b_1 + a_4 b_2. \qquad (21\text{-}25)$$

Eqn. 21-22, for the elements of the inverse matrix, becomes the set

$$b_1 = a_4/D,$$
$$b_2 = -a_2/D,$$
$$b_3 = -a_3/D,$$

and
$$b_4 = a_1/D, \qquad (21\text{-}26)$$

where $D = a_1 a_4 - a_2 a_3$.

The program shown in Fig. 21-1 evaluates the matrix product of two 2×2 matrices,

388

Figure 21-1.   Flow chart for a program to
evaluate matrix products and find the
inverse of a matrix.

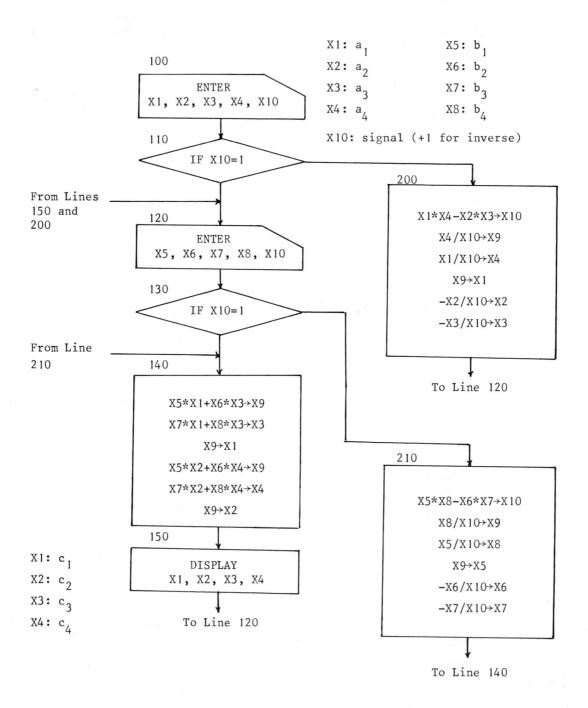

$\underline{\underline{C}}=\underline{\underline{B}}\cdot\underline{\underline{A}}$. It is also possible, within the program, to replace $\underline{\underline{A}}$ by its inverse and $\underline{\underline{B}}$ by its inverse in the product. The matrices are entered in order from right to left, as they appear in the product. $\underline{\underline{A}}$ is entered first, at line 100, with a_1 in X1, a_2 in X2, a_3 in X3, and a_4 in X4. If the inverse of $\underline{\underline{A}}$ is required, the number +1 is entered into X10. The machine then goes to line 200 where the elements of $\underline{\underline{A}}$ are replaced by the elements of $\underline{\underline{A}}^{-1}$. These are calculated using Eqn. 21-26. After the replacement, the machine returns to line 120 to continue executing the main part of the program.

At line 120, the elements of $\underline{\underline{B}}$ are entered, with b_1 in X5, b_2 in X6, b_3 in X7, and b_4 in X8. Entry of +1 into X10 causes the machine to go to line 210, where the elements of $\underline{\underline{B}}$ are replaced by the elements of $\underline{\underline{B}}^{-1}$. After the replacement, the machine returns to line 140 to perform the matrix multiplication.

At line 140, Eqn. 21-24 is used to evaluate the matrix product $\underline{\underline{B}}\cdot\underline{\underline{A}}$ where, of course, $\underline{\underline{A}}$ and $\underline{\underline{B}}$ might be the inverses of the matrices entered. The elements of the product are displayed and the machine returns to line 120 to accept a new matrix. The new matrix then multiplies, from the left, the product $\underline{\underline{B}}\cdot\underline{\underline{A}}$.

Matrix multiplication is carried out at line 140 by first calculating c_1 and storing the result in X9 until c_3 is calculated. Then the value of c_1 is transferred to X1. This sequence must be followed since a_1 is used in the calculation of c_3 and X1 must contain a_1 and not c_1 when c_3 is calculated. Similarly, after it is calculated, c_2 is temporarily stored in X9 until c_4 is calculated, then it is transferred to X2.

The first instruction in the sequence which finds the inverse matrix is for the evaluation of the determinant. Then the first element is computed and stored in X9 until the fourth element is calculated. Finally, the second and third elements are computed.

The order of entry of the matrices should be noted. If the product $\underline{\underline{C}}\cdot\underline{\underline{B}}\cdot\underline{\underline{A}}$ is to be found, $\underline{\underline{A}}$ is entered first, $\underline{\underline{B}}$ is entered next, at line 120, and $\underline{\underline{C}}$ is entered last, when the machine returns to line 120.

The program can also be used to multiply a column matrix by a square matrix. To evaluate

$$\begin{pmatrix} b_1 & b_2 \\ b_3 & b_4 \end{pmatrix} \cdot \begin{pmatrix} a_1 \\ a_3 \end{pmatrix},$$

enter

$$\begin{pmatrix} a_1 & 0 \\ a_3 & 0 \end{pmatrix}$$

into $\underline{\underline{A}}$. The result of the multiplication is the 2×2 matrix

$$\begin{pmatrix} b_1 a_1 + b_2 a_3 & 0 \\ b_3 a_1 + b_4 a_3 & 0 \end{pmatrix},$$

the first column of which is the desired column matrix. Notice that the second element of the original column matrix is entered into the third position of the 2×2 matrix $\underline{\underline{A}}$. That element is labelled a_3 to remind you.

21.2   Matrix Exercises

The following exercises are designed to give you some practice in using the program of Fig. 21-1 and to demonstrate some of the properties of matrix multiplication discussed in section 21.1.

Take great care with the order in which the matrices appear in the statements of the problems and the order in which they must be entered into the program. Also take care with the order in which elements of a given matrix are entered.

Some of the exercises ask for the product of three matrices. For some, intermediate results must be written down and reentered. For others, the program returns the result of the first multiplication in the proper storage locations for the second multiplication.

For the exercises, take

$$\underline{\underline{A}}_1 = \begin{pmatrix} 7.6 & 3.2 \\ 1.4 & 9.8 \end{pmatrix},$$

$$\underline{\underline{A}}_2 = \begin{pmatrix} 2.7 & 6.1 \\ 7.8 & 3.9 \end{pmatrix},$$

$$\underline{\underline{A}}_3 = \begin{pmatrix} 7.6 & -1.9 \\ 8.3 & -1.1 \end{pmatrix},$$

and

$$\underline{\underline{B}} = \begin{pmatrix} 4.7 \\ 3.2 \end{pmatrix}.$$

1. Evaluate the product $\underline{\underline{A}}_1 \cdot \underline{\underline{A}}_2$.

2. Evaluate the product $\underline{\underline{A}}_2 \cdot \underline{\underline{A}}_1$.

3. Evaluate the product $\underline{\underline{A}}_1 \cdot (\underline{\underline{A}}_2 \cdot \underline{\underline{A}}_3)$, then evaluate the product $(\underline{\underline{A}}_1 \cdot \underline{\underline{A}}_2) \cdot \underline{\underline{A}}_3$ and compare the two results. In each case, carry out the multiplication indicated in the parentheses first.

4. Find the inverse to $\underline{\underline{A}}_1$.

5. Evaluate the product $\underline{\underline{A}}_1^{-1} \cdot \underline{\underline{A}}_1 \cdot \underline{\underline{A}}_2$ and compare the answer with $\underline{\underline{A}}_2$.

6. Evaluate the product $\underline{\underline{A}}_1 \cdot \underline{\underline{A}}_2 \cdot \underline{\underline{A}}_1^{-1}$ and compare the answer with $\underline{\underline{A}}_2$.

7. Evaluate the product $\underline{\underline{A}}_1 \cdot \underline{\underline{B}}$.

8. Evaluate the product $\underline{\underline{A}}_1^{-1} \cdot \underline{\underline{A}}_1 \cdot \underline{\underline{B}}$ and compare the answer with $\underline{\underline{B}}$.

9. Use the technique of matrix inversion to solve the equation $\underline{\underline{A}}_1 \cdot \underline{X} = \underline{\underline{B}}$ for the unknown column matrix $\underline{X}$. When $\underline{X}$ has been found, substitute it into $\underline{\underline{A}}_1 \cdot \underline{X}$, evaluate the matrix product, and compare the result with $\underline{\underline{B}}$.

392

10. Use the technique of matrix inversion to solve the equation $\underline{\underline{A}}_1 \cdot \underline{\underline{C}} = \underline{\underline{A}}_2$ for the unknown 2×2 matrix $\underline{\underline{C}}$. When $\underline{\underline{C}}$ has been found, substitute it into $\underline{\underline{A}}_1 \cdot \underline{\underline{C}}$ and compare the result with $\underline{\underline{A}}_2$.

11. Use the technique of matrix inversion to solve the equation $\underline{\underline{C}} \cdot \underline{\underline{A}}_1 = \underline{\underline{A}}_2$ for the unknown 2×2 matrix $\underline{\underline{C}}$. When $\underline{\underline{C}}$ has been found, substitute it into $\underline{\underline{C}} \cdot \underline{\underline{A}}_1$ and compare the result with $\underline{\underline{A}}_2$. Notice that different solutions are obtained for this exercise and exercise 10.

12. Use the technique of matrix inversion to solve the equation $\underline{\underline{A}}_1 \cdot \underline{\underline{C}} \cdot \underline{\underline{A}}_2 = \underline{\underline{A}}_3$ for the unknown 2×2 matrix $\underline{\underline{C}}$. When $\underline{\underline{C}}$ has been found, substitute it into $\underline{\underline{A}}_1 \cdot \underline{\underline{C}} \cdot \underline{\underline{A}}_2$ and compare the result with $\underline{\underline{A}}_3$.

21.3   Lens Systems

Systems of thin lenses and their influence on paraxial rays are considered. The techniques presented here, however, can easily be extended to thick lenses.

In general, the lenses we consider are aligned along the x axis and their positions and focal lengths are given. If a lens is at x_1 then, because the lens is thin, refraction is assumed to take place at the plane through x_1, perpendicular to the x axis. Fig. 21-2 is a diagram showing the refraction of a paraxial ray by a single thin lens. Only the plane through the center of the lens is shown, at $x = x_1$.

Figure 21-2.   A paraxial ray
refracted at a thin lens.
The sizes of the angles are
exaggerated.

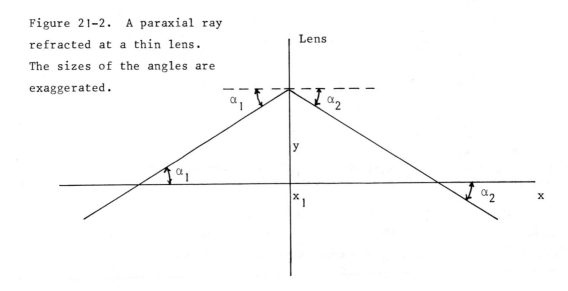

Angles are shown much larger than they actually are for paraxial rays. They are measured from a line parallel to the x axis and are positive in the counterclockwise direction, as viewed in the diagram. The incident ray makes the angle α_1 with the x axis and the refracted ray makes the angle α_2 (a negative angle for the ray shown).

It is convenient to carry out the calculation with the angles measured in radians. The conversion to degrees, if desired, is accomplished by multiplying the angle in radians by 180 and dividing the result by π.

The object might be anywhere along the incident ray and the image might be anywhere along the refracted ray in the diagram. At first, we consider point sources on the x axis. The object is a distance o and the image is a distance i from the lens. These quantities are positive for real objects and images, respectively, and negative for virtual objects and images, respectively.

For purposes of tracing a ray, the two quantities of interest are the angle α it makes with the x axis and the distance y from the x axis to the point where it strikes the lens. A column matrix, called the ray matrix and denoted by $\underline{\underline{r}}$, is formed by these two quantities:

$$\underline{\underline{r}} = \begin{pmatrix} \alpha \\ y \end{pmatrix}. \tag{21-27}$$

Two different square matrices are used. The first describes the influence of a lens on the ray and the second describes the change in $\underline{r}$ as the ray travels between lenses. The first matrix, called the refraction matrix, multiplies the ray matrix for the incident ray and produces the ray matrix for the refracted ray. The second matrix, called the transfer matrix, multiplies the ray matrix for the ray as it leaves the lens and produces the ray matrix for the ray as it encounters the next lens.

Successive multiplications by refraction and transfer matrices trace the ray through the lens system.

394

The refraction matrix for a given lens can be developed from the lens equation

$$\frac{1}{o} + \frac{1}{i} = \frac{1}{f} .$$

(21-28)

Here f is the focal length of the lens, positive if the lens causes parallel incident rays to converge behind the lens and negative if it causes them to diverge. Since the angles α_1 and α_2 are measured in radians and the rays are paraxial,

$$o = y/\alpha_1$$

(21-29)

and

$$i = -y/\alpha_2.$$

(21-30)

To obtain these results, use $\tan\alpha_1 = y/o$ and $\tan\alpha_2 = -y/i$, then approximate the tangents by the angles. For small angles, these approximations are valid provided the angles are measured in radians.

Upon substitution of Eqns. 21-29 and 21-30, Eqn. 21-28 becomes

$$\frac{\alpha_1}{y} - \frac{\alpha_2}{y} = \frac{1}{f}$$

(21-31)

or

$$\alpha_2 = \alpha_1 - y/f.$$

(21-32)

This is one equation in a set of two which are to be reformulated in matrix notation. The other is extremely simple. Let y_1 be the height at which the incident ray strikes the lens and let y_2 be the height at which the refracted ray leaves the lens. Since the lens is thin, $y_2 = y_1$. Now write the two equations, $\alpha_2 = \alpha_1 - y_1/f$ and $y_2 = y_1$ in matrix notation:

$$\begin{pmatrix} \alpha_2 \\ y_2 \end{pmatrix} = \begin{pmatrix} 1 & -1/f \\ 0 & 1 \end{pmatrix} \cdot \begin{pmatrix} \alpha_1 \\ y_1 \end{pmatrix}.$$

(21-33)

The matrix

$$\underline{\underline{R}} = \begin{pmatrix} 1 & -1/f \\ 0 & 1 \end{pmatrix} \qquad (21\text{-}34)$$

is the refraction matrix for the lens. Notice that this matrix does its intended job. Multiplication by the refraction matrix transforms the ray matrix for the incident ray into the ray matrix for the refracted ray. In matrix notation

$$\underline{\underline{r}}_2 = \underline{\underline{R}} \cdot \underline{\underline{r}}_1 \qquad (21\text{-}35)$$

where $\underline{\underline{r}}_1$ is the ray matrix for the incident ray and $\underline{\underline{r}}_2$ is the ray matrix for the refracted ray.

If a second lens is present, the ray leaving the first lens is the ray incident on the second lens. The ray matrix which describes the ray when it strikes the second lens is denoted by $\underline{\underline{r}}_3$:

$$\underline{\underline{r}}_3 = \begin{pmatrix} \alpha_3 \\ y_3 \end{pmatrix}. \qquad (21\text{-}36)$$

The direction of the ray does not change between the lenses but the distance to the ray from the optic axis does. In fact

$$y_3 = y_2 + d\,\tan\alpha_2 \simeq y_2 + d\alpha_2$$

where d is the distance between lenses. The two equations $\alpha_3 = \alpha_2$ and $y_3 = y_2 + d\alpha_2$, can be written in matrix form:

$$\begin{pmatrix} \alpha_3 \\ y_3 \end{pmatrix} = \begin{pmatrix} 1 & 0 \\ d & 1 \end{pmatrix} \cdot \begin{pmatrix} \alpha_2 \\ y_2 \end{pmatrix}. \qquad (21\text{-}37)$$

396

The matrix

$$\underline{\underline{T}} = \begin{pmatrix} 1 & 0 \\ d & 1 \end{pmatrix} \tag{21-38}$$

is called the transfer matrix and multiplication by it changes the ray matrix for the ray leaving the first lens into the ray matrix for the same ray when it strikes the second lens:

$$\underline{r}_3 = \underline{\underline{T}} \cdot \underline{r}_2. \tag{21-39}$$

A transfer matrix can also be used to find the ray matrix at the first lens, given the position of the source and the inital direction of the ray. If the source is at x_0, y_0 and the ray leaves the source at angle α, then the inital ray matrix is

$$\underline{r} = \begin{pmatrix} \alpha \\ y_0 \end{pmatrix} \tag{21-40}$$

and the appropriate transfer matrix is

$$\underline{\underline{T}} = \begin{pmatrix} 1 & 0 \\ x_1 - x_0 & 1 \end{pmatrix} \tag{21-41}$$

where x_1 is the position of the first lens. The product $\underline{\underline{T}} \cdot \underline{r}$ is the ray matrix for the ray as it strikes the first lens.

Refraction and transfer matrices can be used to follow a ray through a lens system. Number the lenses 1, 2, 3, ..., in order along the optic axis. Lens 1 is struck by the ray first. For each lens, write the refraction matrix. Let $\underline{\underline{R}}_i$ be the matrix for lens i. Number the regions between the lenses 1, 2, 3, ..., with region 1 following lens 1 and, in general, region i following lens i. If $\underline{r}_1$ is the ray matrix for the ray as it leaves the source and $\underline{r}_2$ is the ray matrix as it emerges from the last lens, then

$$\underline{\underline{r}}_2 = \cdots \underline{\underline{R}}_3 \cdot \underline{\underline{T}}_2 \cdot \underline{\underline{R}}_2 \cdot \underline{\underline{T}}_1 \cdot \underline{\underline{R}}_1 \cdot \underline{\underline{T}}_0 \cdot \underline{\underline{r}}_1 . \qquad (21\text{-}42)$$

Read this equation from right to left, in the order of matrix multiplication. The ray travels from the source to the first lens, it is refracted by this lens, then it traverses the space between lens 1 and lens 2. It is refracted by lens 2 and traverses the space between lens 2 and lens 3. Its progress continues until it is refracted at the last lens of the system.

Sometimes the intersection of the final ray with the optic axis is required. This can be found by using

$$x = x_L - y_L/\tan\alpha_L \simeq x_L - y_L/\alpha_L , \qquad (21\text{-}43)$$

where x_L is the position of the last lens, y_L is the y coordinate of the point where the ray leaves the last lens, and α_L is the angle the ray makes with the x axis as it leaves the last lens. The distance $-y_L/\alpha_L$ of the intersection point from the last lens can be calculated quite easily with the program of Fig. 21-1. After the ray matrix for the emerging ray has been found, simply evaluate $-X3/X1$. The result is positive for a point behind the lens and negative for a point in front.

The intersection of a ray with a plane perpendicular to the optic axis can be found quite easily by applying another transfer matrix, this time carrying the ray from the last lens to the plane. The distance which appears in the third element of the matrix is the distance between the last lens and the plane. After multiplication, the lower entry in the ray matrix gives the y coordinate of the intersection. Sometimes it is convenient to insert this last transfer matrix solely for the purpose of finding another point on the ray after it emerges from the last lens. This is useful for drawing a ray diagram.

As part of the solution to problem 8 of section 20.3, you calculated image distances for sources in front of a thin lens. The results can be used to check the matrix technique and the program of Fig. 21.1.

Problem 1. The first surface of a thin, double convex lens has a radius of 0.2 m and the second surface has a radius of 0.6 m. The glass has an index of refraction of 1.47 and is in air (n=1).

a. Use Eqn. 20-26 to find the focal length of the lens. This repeats part a of problem 8 in section 20.3.

b. A point source is located on the optic axis, 2 m in front of the lens. Consider the ray which leaves the source 0.001° above the axis. Construct the transfer matrix which carries the ray from the source to the lens and the refraction matrix for the lens. Use these matrices and the program of Fig. 21-1 to find the ray matrix for the ray as it leaves the lens. Finally, find the point where the ray crosses the optic axis and compare the result with the thin lens image distance found in answer to problem 8, section 20.3. Don't forget to convert the given angle to radian measure.

c. Repeat the calculation of part b for a point source on the axis, 0.5 m in front of the lens.

d. Repeat the calculation of part b for a point source on the axis, 0.1 m in front of the lens.

If two paraxial rays emerge from the last lens, one at angle α_1 and distance y_1 from the axis and the other at angle α_2 and distance y_2 from the axis, they intersect at

$$x = -(y_2-y_1)/(\alpha_2-\alpha_1), \qquad (21-44)$$

$$y = (y_1\alpha_2-y_2\alpha_1)/(\alpha_2-\alpha_1). \qquad (21-45)$$

Here x is measured from the position of the last lens and is positive if the intersection is behind the lens, negative if the intersection is in front of the lens. y is positive for points above the axis, negative for points below the axis. These expressions can be derived quite easily, using some trigonometry and the approximation $\tan\alpha\approx\alpha$, valid for small angles, measured in radians. They are useful for finding the position of an image.

Problem 2. A 0.001 m arrow is placed 0.35 m in front of a thin lens with a

0.12 m focal length. The arrow is perpendicular to the optic axis and its tail rests on that axis.

a. Find the position of the image of the tail by tracing the ray with $\alpha=0.001^{\circ}$ and finding its intersection with the optic axis after refraction. Find the appropriate transfer and refraction matrices and use the program of Fig. 21-1.

b. Find the position of the image of the head by tracing rwo rays, one with $\alpha=0$ and the other with $\alpha=-0.001^{\circ}$, and finding their intersection with each other.

c. Find the position of the image of the head by tracing the ray which leaves the head with $\alpha=0$, then finding the intersection of the refracted ray with the plane perpendicular to the optic axis at the position of the image of the tail. Two transfer matrices are required.

d. Compute the magnification as the ratio of the image length to the object length and compare the result with $-i/o$.

Before going on to the study of optical instruments, we use matrix techniques to demonstrate some important properties of thin lenses. These properties are useful for the understanding of the optical instruments studied later.

The region in front of a thin converging lens, with focal length f, is conveniently divided into three zones. One zone is more than 2f in front of the lens and, if the object is in this zone, the lens produces a real, inverted image, which is smaller than the object. The second zone is between 2f and f in front of the lens and, for objects in this zone, the lens produces a real, inverted image which is larger than the object. The closest zone lies between the object focus and the lens. For objects in this zone, the lens produces a virtual, erect image which is larger than the object.

A thin lens with a negative focal length produces a virtual, erect image which is smaller than the object, no matter where the object is placed in front of the lens.

We explicitly verify these statements for two lenses, one with positive and one with negative focal length.

Problem 3.

a.   Consider a 0.005 m long arrow, placed in front of a 0.18 m focal length thin lens. The arrow is perpendicular to the optic axis and its tail rests on that axis. For each of the following object positions, use the program of Fig. 21-1 to find the position and length of the image: o=0.54 m (3f), 0.36 m (2f), 0.27 m (1.5f), and 0.09 m (0.5f). First find the image of the tail by tracing a paraxial ray and finding where it intersects the optic axis. Then trace a paraxial ray from the head and find where it intersects the plane perpendicular to the axis at the image of the tail. This is the technique which was used in part c of problem 2. In each case, calculate the magnification.

b.   Repeat the calculation of part a, but take the focal length to be -0.18 m.

We now use matrix techniques to investigate some characteristics of telescopes, microscopes, and camera lenses. In practice, these instruments are constructed with thick lenses and the lens system is usually quite complicated. For an introduction, however, the examples used here deal with thin lenses and the simplest possible systems.

The purpose of a telescope is to gather light from objects which are far away and to create an image which is larger than the object appears to be when viewed from the same distance, but without the aid of the telescope.

Two lenses are required. The first lens, which is positive, is called the objective lens. It is usually large in diameter in order to collect as much light as possible and thus produce a bright image. Since the object is far away, light rays which enter the telescope are nearly parallel to the optic axis and the image is near the image focal plane of the objective lens. This image is, of course, real, inverted, and smaller than the object.

The image formed by the objective lens is the object for the second lens, called the ocular or eyepiece. The primary purpose of this lens is to magnify the image of the objective.

There is a secondary purpose, which arises from the need to relieve eye strain. Since the eye, in a relaxed state, focuses objects which are far away, there is less strain if the final image of the telescope is far from the eye and one purpose of the eyepiece is to produce such an image.

To accomplish these goals, the eyepiece is usually a lens with a short, positive focal length and it is placed so that the image formed by the objective is at the front focal plane of the eyepiece. It is usual to design a telescope so that the distance between lenses is the sum of the focal lengths. The image focal plane of the objective and the object focal plane of the eyepiece then coincide. Since the image formed by the objective is always at or just behind the image focal plane of that lens, this means that it is always at or just behind the object focal plane of the eyepiece.

The final image is then virtual and inverted. The image formed by the objective is inverted and the eyepiece does not change that orientation. Check these statements by referring to problem 3.

Telescopes of the type described above are called astronomical telescopes. The final image is inverted, but this makes no difference when viewing stars and planets. The image can be made erect by the addition of a third lens or a prism and the instrument is then called a terrestrial telescope.

We consider astronomical telescopes, constructed with two positive lenses. For a given pair of lenses, the closer the image of the objective lens is to the focal plane of the eyepiece, the greater the magnification. If the image is exactly at the focal plane, the rays emerging from the eyepiece are parallel and the image formed by that lens is said to be at infinity.

If $\underline{\underline{r}}_i$ is the ray matrix for the incident ray and $\underline{\underline{r}}_e$ is the ray matrix for the emerging ray, then, for the two lens telescope,

$$\underline{\underline{r}}_e = \underline{\underline{R}}_2 \cdot \underline{\underline{T}}_1 \cdot \underline{\underline{R}}_1 \cdot \underline{\underline{T}}_0 \cdot \underline{\underline{r}}_i \qquad (21\text{-}46)$$

where

$$\underline{\underline{T}}_0 = \begin{pmatrix} 1 & 0 \\ d_1 & 1 \end{pmatrix} ,$$

$$\underline{\underline{R}}_1 = \begin{pmatrix} 1 & -1/f_{ob} \\ 0 & 1 \end{pmatrix} ,$$

$$\underline{\underline{T}}_2 = \begin{pmatrix} 1 & 0 \\ d_2 & 1 \end{pmatrix} ,$$

and

$$\underline{\underline{R}}_2 = \begin{pmatrix} 1 & -1/f_{ey} \\ 0 & 1 \end{pmatrix} .$$

Here d_1=distance from object to objective lens, d_2=distance between lenses, f_{ob}=focal length of the objective lens, and f_{ey}=focal length of the eyepiece. The matrix product $\underline{\underline{S}}=\underline{\underline{R}}_2\cdot\underline{\underline{T}}_1\cdot\underline{\underline{R}}_1$ is the system matrix and it can be determined once for a given telescope and then, for any incident ray, it multiplies $\underline{\underline{T}}_0\cdot\underline{r}_i$ to give the emerging ray. It need not be recomputed for different incident rays. $\underline{\underline{T}}_0\cdot\underline{r}_i$, of course, is different for different incident rays.

Problem 4. As astronomical telescope consists of two lenses: an objective with a focal length of 0.6 m and an eyepiece with a focal length of 0.12 m. The lenses are initially separated by 0.7 m and used to view an arrow 0.1 m long, placed perpendicular to the optic axis, 200 m in front of the first lens. Notice that the front focal point of the eyepiece is between the first lens and its back focal point.

a. Use the program of Fig. 21-1 to find the position of the final image and the magnification of the telescope. First find the system matrix for the telescope. Then consider a single paraxial ray from the tail of the arrow, assumed to be on the optic axis, and find its intersection with the axis. This locates the plane of the image. Finally, consider a paraxial ray from the head of the arrow and find its intersection with the plane of the image.

b. Now move the eyepiece farther from the objective lens so that the distance
    between lenses is 0.72 m and repeat the calculations of part a. Compare the
    magnification with that found in part a. The lenses are now separated by
    a distance which is equal to the sum of the focal lengths.

c. Again move the eyepiece, this time so that the distance between lenses is
    0.74 m. Repeat the calculations of part a. Compare the magnification with
    that found in parts a and b.

Notice that the magnification changes dramatically with the distance
between lenses. For most telescopes, the distance between lenses is adjustable
over a range of a few millimeters rather than a few centimeters as in the example.
For the last case considered, the image is behind the telescope and it is hard
for the viewer's eye to focus on it.

Notice also that, in each case, the image is smaller than the object. The
magnification does not give a direct measure of how well the instrument does the
job for which it is intended. It compares image and object sizes. What is needed
is a comparison of the image size with the apparent size of the object as viewed
without the telescope. A quantity called the angular magnification has been
devised to provide this comparison.

The angular magnification is computed as follows. An arrow object is viewed
and the ray from the head to the center of the objective lens is considered. Since
the central ray from the tail goes straight through the telescope, the angle α_i
between the ray from the head and the optic axis is a measure of the angle subtended
by the arrow at the unaided eye. In principle, the ray considered should go from
the head of the arrow to the center of the eye but, since the telescope is far
shorter than the object distance, this ray is, for all practical purposes, in the
same direction as the one to the objective lens center. The ray to the lens center
is now traced through the telescope and the angle α_e which it makes with the axis
on emergence is calculated. This angle is the angle subtended at the eye by the
final image formed by the telescope. The ratio α_e/α_i is the angular magnification.

At the objective lens, the ray matrix for the incident ray used to find
the angular magnification is

$$\begin{pmatrix} \alpha_i \\ 0 \end{pmatrix}$$

where $\alpha_i = -\ell/s$. Here ℓ is the object's length and s is its distance from the telescope. The system matrix is used to find α_e and the ratio α_e/α_i is computed.

Problem 5.   Consider the telescope of problem 4, adjusted to that the distance between lenses is equal to the sum of the focal lengths.

a.   Consider a ray which enters the objective lens at its center and at an angle of 3°. Take this to be α_i. Use the program of Fig. 21-1 to find where the ray enters the second lens and where the emerging ray crosses the plane 0.5 m behind the eyepiece. On a piece of graph paper, make a scale drawing of the lens system and show the ray.

b.   Compute the angular magnification.

c.   Suppose an arrow is 10 m in front of the objective lens, with its tail on the optic axis, and suppose it subtends an angle of 3° at the objective lens. Find the position of the final image by tracing a paraxial ray from the tail and finding where it crosses the optic axis. On the drawing of part a, draw the image. It extends from the axis to the ray drawn in part a, extended backward into the region between the lenses.

d.   Use $\alpha_i = -\ell/s$ to find the length of the object arrow.

e.   Calculate the linear magnification of the telescope as the ratio of the image length to the object length.

Typically, the linear magnification of a telescope is less than 1 and the image is smaller than the object. The angular magnification, however, is greater than 1 and the image is larger than the object as seen by the unaided eye, at the same distance.

We now turn to a discussion of microscopes, instruments which produce images which are much larger than the objects themselves. A compound microscope consists of an objective lens and an eyepiece, just like a telescope. The purposes of the two instruments, however, are quite different and the difference leads to the use of lenses with different focal lengths and to different relative placement of the object and the lenses.

For this introductory discussion, we consider a microscope composed of a single positive lens for the objective and a single positive lens for the eyepiece. The object is placed just in front of the front focal point of the objective lens. This lens forms a real, inverted, and magnified image well beyond the back focal point.

The microscope is designed for a selected object distance. The distance from the back focal point of the objective lens to the image formed by that lens is called the tube length of the microscope and it is fixed once the object distance and the focal length of the objective lens are selected. Microscopes are usually designed with a standard tube length of 0.16 m.

The eyepiece is placed so that its front focal point is at the image formed by the objective lens. The distance between lenses is then the sum of the two focal lengths and the tube length. The intermediate image is the object for the eyepiece and, since it is at the front focal plane, the final image is at infinity.

The tube length L plays an important role in the determination of the magnification. The magnification of the objective lens can be calculated using $m=-i/o$ and, since $1/o=(1/f_{ob})-(1/i)$, $m=1-(i/f_{ob})$. Now $i=f_{ob}+L$, so $m=-L/f_{ob}$.

The eyepiece is characterized by its angular magnification. This is defined so that it is a comparison of the angle subtended by the final image at the eye and the angle that would be subtended by the intermediate image at the eye if it were viewed without the benefit of the eyepiece. The intermediate image is used, of course, since this is the object for the eyepiece. To calculate the angle subtended in the latter case, the eye is placed, not at the end of the microscope tube, but rather, as close as possible to the imtermediate image and still allow the eye to focus the rays. Although this distance varies from eye to eye, we take it to be 0.25 m, an average distance. The angle subtended by the intermediate image at an eye 0.25 m away is $y_i/0.25$ where y_i is the length of the intermediate image in meters. To calculate the angle subtended by the final image when the eyepiece is used, we consider the ray from the head of the intermediate image, through the center of the eyepiece. Since the intermediate image is at the focal plane of the eyepiece, it is f_{ey} distant from the lens and the angle between the ray and the optic axis is y_i/f_{ey}. This ray is undeviated by the eyepiece, so y_i/f_{ey} also gives the angle subtended by the final image at the eye. The

406

angular magnification of the eyepiece is therefore $(y_i/f_{ey})/(y_i/0.25)=0.25/f_{ey}$, where f_{ey} is in meters.

The magnification of the microscope as a whole is the product of the linear magnification of the objective and the angular magnification of the eyepiece. That is

$$m = -(L/f_{ob})(0.25/f_{ey}), \qquad\qquad (21\text{-}47)$$

where all distances are in meters.

Problem 6.   The objective lens of a microscope has a focal length of 7×10^{-3} m and the eyepiece has a focal length of 27×10^{-3} m. The microscope is designed so that when the object is 7.4×10^{-3} m in front of the objective lens, the final image is at infinity.

a.   Use the program of Fig. 21-1 to trace a paraxial ray from an object on the axis and find the position of the intermediate image. What is the tube length?

b.   The eyepiece is placed so that its front focal point is at the image formed by the objective lens. How far is the eyepiece from the objective?

c.   On a piece of graph paper, draw the optic axis and mark, to scale, the positions of the lenses, their front and back focal points, and the positions of the object and intermediate image. Label the tube length on the diagram.

d.   An arrow, 5×10^{-4} m long, is placed at the design object position. It is perpendicular to the optic axis and its tail rests on that axis. Trace the ray from the head through the center of the objective lens. Find where it crosses the plane of the intermediate image and verify that the image is real, inverted, and magnified. Continue to trace the ray and find its direction when it emerges from the eyepiece.

e.   Calculate the magnification of the objective lens by evaluating the ratio of the lengths of the intermediate image and the object arrow.

f.   Calculate the angular magnification of the eyepiece. To do this, consider the ray from the head of the intermediate image through the center of the eyepiece and find the angle it makes with the optic axis. Then evaluate the ratio of this angle to the angle y/0.25, where y is the length of the intermediate image in meters.

g.   What is the magnification of the microscope? Take the product of the results
     of parts e and f. Now calculate the magnification another way. If the object
     were viewed by an unaided eye 0.25 m away, it would subtend an angle of
     $5 \times 10^{-4}/0.25 = 2 \times 10^{-3}$ radians. Calculate the ratio of the angle found in part d
     to this angle.

     A lens for a camera must produce a real image. There are many lens
configurations which do this, of course. As an example, we simply use a single
positive lens. Regardless of the object distance, the image formed by the lens
must be in the plane of the film and this is accomplished, for various object distances, by moving the lens toward or away from the film. The following problem is
designed to help you understand how a camera lens works.

Problem 7.   The lens of a camera has a focal length of 50×10^{-3} m. We shall find
the lens-to-film distance for various object distances.

a.   As the lens moves outward, for example, the object distance decreases and the
     lens-to-film distance increases. Let d be the distance from the object,
     assumed to be on the optic axis, to the film and suppose the image is in
     the plane of the film. Then d=o+i. Substitute o=d-i into $(1/o)+(1/i)=(1/f)$
     and solve for i. In particular, show that

$$i = (d/2) \left[1 - (1-4f/d)^{\frac{1}{2}} \right] .$$

b.   Find the lens-to-film distance i for each of the following values of d: 0.5 m,
     1 m, 2 m, 10 m, and 100 m.

     Notice that the lens must be moved further when the object moves from 0.5 m
to 1 m than it does when the object moves from 1 m to 100 m. If you have a
50 mm lens for an SLR camera, look at the barrel markings which tell how much
the lens should be turned to focus objects at various distances. They are
far apart for small distances and close together for large distances. As
the lens turns, of course, it is moving along a screw thread and the lens-to-
film distance is changing.

c. Suppose a point source is on the optic axis at the position for which d=0.5 m. The camera is focused. Trace a paraxial ray from the source and find where it crosses the optic axis again. Verify that this is the image distance as calculated in part b.

d. Repeat the calculation of part c for d=2 m.

e. Consider an arrow of length ℓ, in front of the camera. The ray from its head to the lens center makes an angle of ℓ/o radians with the optic axis. If there were no lens, the object would subtend the angle ℓ/d, so the angular magnification is $(\ell/o)/(\ell/d)=d/o$. Evaluate d/o for each of the values of d given in part b.

A zoom lens for a camera is one whose angular magnification can be changed without changing the object-to-image distance. The image can be enlarged and still remain in focus at the plane of the film. There are many ways to accomplish this job. While most zoom lenses consist of a complex grouping of lenses, we consider a simple system consisting of three positive lenses. The angular magnification is changed by moving the middle lens relative to the other two. To retain the image in the plane of the film, the back lens must also move. For the first problem, however, the back lens is kept stationary.

Problem 8. A zoom lens consists of a front and back lens, each with focal length 0.05 m. They are 0.3 m apart. The center lens has a focal length of 0.1 m and it is positioned a distance d from the front lens. Consider a point source on the optic axis, 50 m in front of the front lens. For each of the following values of d, trace a paraxial ray from the source and find the position of the image. Also, for each value of d, trace the ray which enters the center of the front lens at an angle of -0.001 radians and find the angular magnification. If α_e is the angle the emerging ray makes with the optic axis, then the angular magnification is closely approximated by $-\alpha_e/0.001$. Take d=0.06 m, 0.1 m, 0.14 m, and 0.18 m. The distance between the center and back lenses is 0.3-d if d is in meters.

Notice the wide range of angular magnifications which can be achieved. For the conditions of the last problem, the image does not quite remain in focus at the plane of the film, however. This defect can be corrected by moving the back lens. The analysis is straightforward and is carried out in the next problem.

Problem 9. Suppose the point object is on the optic axis, a distance d_0 in front of the front lens, and the film is a distance ℓ behind the front lens. These distances are fixed. Take d_1 to be the distance between the front and center lenses, d_2 to be the distance between the center and back lenses, and d_3 to be the distance between the back lens and the film. These distances change as the center and back lenses are moved. Consider a ray which enters the front lens at the angle α_i.

a. Let

$$\begin{pmatrix} \alpha \\ y \end{pmatrix}$$

be the ray matrix for the ray as it leaves the center lens. It can be found by carrying out the matrix multiplication

$$\begin{pmatrix} \alpha \\ y \end{pmatrix} = \begin{pmatrix} 1 & -1/f_2 \\ 0 & 1 \end{pmatrix} \cdot \begin{pmatrix} 1 & 0 \\ d_1 & 1 \end{pmatrix} \cdot \begin{pmatrix} 1 & -1/f_1 \\ 0 & 1 \end{pmatrix} \cdot \begin{pmatrix} 1 & 0 \\ d_0 & 1 \end{pmatrix} \cdot \begin{pmatrix} \alpha_i \\ 0 \end{pmatrix}.$$

The ray matrix for the ray as it strikes the film is given by

$$\begin{pmatrix} \alpha_e \\ y_e \end{pmatrix} = \begin{pmatrix} 1 & 0 \\ d_3 & 1 \end{pmatrix} \cdot \begin{pmatrix} 1 & -1/f_3 \\ 0 & 1 \end{pmatrix} \cdot \begin{pmatrix} 1 & 0 \\ d_2 & 1 \end{pmatrix} \cdot \begin{pmatrix} \alpha \\ y \end{pmatrix}.$$

The last matrix in the product (on the left) is the transfer matrix which carries the ray from the last lens to the film. For various values of d_1, we wish to solve for d_3 when the image is at the plane of the film and $y_e = 0$. Analytically carry out the last matrix multiplication and equate the lower entries on the left and right sides of the equation to show that

$$\alpha d_3 = \alpha d_2 d_3 / f_3 + \alpha d_2 - y d_3 / f_3 + y = 0.$$

Substitute $d_2 = \ell - d_1 - d_3$ and solve for d_3. The result is

$$d_3 = \frac{1}{2\alpha} \left\{ \alpha(\ell - d_1) + y + \left[(\alpha\ell - \alpha d_1 + y)^2 - 4\alpha(\alpha\ell - \alpha d_1 + y) f_3 \right]^{\frac{1}{2}} \right\} .$$

b. Take $\alpha_i = 2 \times 10^{-5}$ radians, $d_0 = 50$ m, $f_1 = f_3 = 0.05$ m, and $f_2 = 0.1$ m. These are the same parameters that were used in problem 8. Take $d_1 = 0.1$ m and use the program of Fig. 21-1 to find the ray matrix for the ray as it leaves the center lens. Then calculate d_3 and use $\ell = d_1 + d_2 + d_3$ to calculate d_2.

c. Repeat the calculation of part b, but this time use $d_1 = 0.18$ m.

d. To see how much the back lens must be moved in order to retain a focused image as the angular magnification is changed, compute $d_1 + d_2$ for the two situations and compare. The angular magnifications are nearly the same as those found in problem 8 for the same values of d_1.

Chapter 22

INTERFERENCE AND DIFFRACTION

The programmable calculator can be used to examine interference and diffraction phenomena. Programs are given which calculate the resultant disturbance when several waves are present simultaneously at the same place. The importance of coherence is demonstrated by considering several waves with randomly selected phases. Huygens' principle is discussed and used to calculate the diffraction patterns of both single and double slits. Changes in the patterns are investigated as the viewing screen is brought closer to the slits and the appearance of the geometric image is demonstrated. This material supplements Chapters 45 and 46 of PHYSICS, Chapters 40 and 41 of FUNDAMENTALS OF PHYSICS, and similar sections of other texts.

22.1   Interference

Many optical phenomena cannot be understood in terms of geometrical optics and, to understand them, the wave nature of light must be taken into account. Among these are interference and diffraction effects, which arise from the superposition of two or more waves in the same region of space.

As described in section 19.1, light consists of an electric and a magnetic field which propagate in space. When two such waves reach the same point at the same time, the total electric field there is the vector sum of the two electric fields and the total magnetic field is the vector sum of the two magnetic fields. Since the intensity or brightness is proportional to the square of the magnitude of the total electric or magnetic field, it can be up to four times the intensity associated with one of the waves or it can vanish. The first situation occurs when the two electric fields have the same magnitude and are in the same direction at the same time. The second situation occurs when the two electric fields have the same magnitude and are in opposite directions at the same time. If a light wave illuminates a certain small region of space, the addition of a second wave may cause the region to brighten or it may cause it to darken, depending on the relative orientations of the fields associated with the waves. To see these effects, precautions must be taken to assure the fields have the desired orientations over periods of time which are long enough to make observations. This is not always easy to do, as we shall see.

411

The effect of wave superposition is called interference. Neither wave, however, changes the other. If the waves cross paths and then go on to separate regions of space, each wave emerges from the region of crossing exactly the same as it would if the other wave never existed. They pass through each other unchanged. What is different because of the presence of a second wave is the total disturbance in the region where the waves exist simultaneously.

While the above discussion deals specifically with light, it should be remarked that interference phenomena occur with all waves, no matter what their nature. Sound and water waves are no exception and examples can be heard in a room with a stereo playing or seen at the beach.

The plotting program of Fig. 2-4 (Appendix C) can be used to study interference. The function to be plotted is proportional to the intensity and it is plotted as a function of position. In general, the intensity is also time dependent. However, for light the frequency is so great that the eye does not respond to the variations with time and the average intensity is a valid measure of the brightness as seen by the eye.

We consider two point sources on the y axis, at $y=-\Delta$ and $y=+\Delta$, respectively. The intensity pattern is viewed on a screen which is placed at the plane $x=D$ as shown in Fig. 22-1.

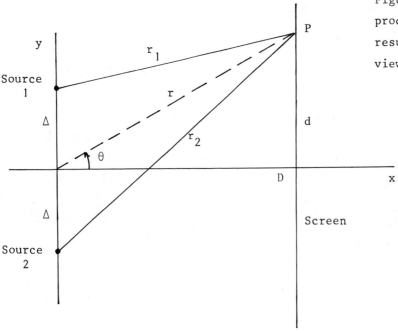

Figure 22-1. Two point sources produce light waves and the resultant intensity pattern is viewed on a screen.

We calculate the average intensity at point P, a distance d from the x axis. Both sources emit sinusoidal waves, with the same amplitude and frequency. At P the disturbance due to source 1 is given by

$$E_1 = \frac{a}{r_1} \sin(kr_1 - \omega t) \qquad (22\text{-}1)$$

and that due to source 2 is given by

$$E_2 = \frac{a}{r_2} \sin(kr_2 - \omega t + \phi). \qquad (22\text{-}2)$$

Here $k = 2\pi/\lambda$ where λ is the wavelength of the light and ω is the angular frequency. As for all waves, ω/k is the speed of the wave. Inclusion of the phase constant ϕ allows us to consider waves which are out of phase at the sources. Angles are in radians.

Strictly speaking, the electric fields are vectors and their directions should be taken into account. When they are summed to find the total field, vector addition should be used. For each wave, the field is perpendicular to the direction of propagation and, as the observation point P is varied, the relative orientation of the two fields changes. These changes in the relative orientation of the fields do influence variations in the intensity from point to point, but this is not a main feature of the phenomena to be demonstrated and, for purposes of the demonstration, we take the fields to be scalar quantities.

The total disturbance at P is given by

$$E = E_1 + E_2 = \frac{a}{r_1} \sin(kr_1 - \omega t) + \frac{a}{r_2} \sin(kr_2 - \omega t + \phi). \qquad (22\text{-}3)$$

This disturbance is sinusoidal in time and has angular frequency ω. The equation for the total disturbance can be written in the form

$$E = E_0 \sin(-\omega t + \phi), \qquad (22\text{-}4)$$

where E_0 depends on a, r_1, r_2, k, and ϕ.

A geometric method can be used to find an expression for E_0. Consider two vectors, one of which has magnitude a/r_1 and makes the angle $kr_1 - \omega t$ with the x axis, while the other has magnitude a/r_2 and makes the angle $kr_2 - \omega t + \phi$ with the x axis. The

414

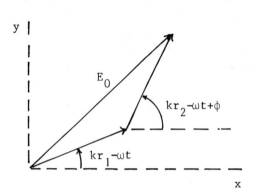

Figure 22-2. The addition of
phasors associated with two
sinusoidal waves.

vectors are shown in Fig. 22-2. Also shown is the resultant, which has magnitude E_0 and makes the angle $-\omega t+\alpha$ with the x axis. Eqn. 22-3 is simply the statement, expressed in mathematical terms, that the y component of the resultant is the sum of the y components of the two vectors in the sum.

The vectors depicted in the diagram are not the electric field vectors. They are constructions used to show the relative phases and amplitudes of the various waves and to aid in the calculation of E_0. They are called phasors.

To find E_0, first sum the x components of the two constituent phasors, then sum the y components, and finally evaluate the square root of the sum of the squares of these two sums. The result, for E_0^2, is

$$E_0^2 = \left[\frac{a}{r_1}\cos(kr_1-\omega t) + \frac{a}{r_2}\cos(kr_2-\omega t+\phi)\right]^2$$

$$+ \left[\frac{a}{r_1}\sin(kr_1-\omega t) + \frac{a}{r_2}\sin(kr_2-\omega t+\phi)\right]^2. \qquad (22-5)$$

This expression can be simplified considerably. If the whole diagram is rotated through any angle, the relative orientation of the phasors is not changed and neither is the magnitude of E_0. Mathematically, the rotation is accomplished by adding the same amount to each of the angles between the phasors and the x axis. We choose to add ωt. Then the equation for E_0^2 becomes

$$E_0{}^2 = \left[\frac{a}{r_1} \cos(kr_1) + \frac{a}{r_2} \cos(kr_2+\phi)\right]^2$$

$$+ \left[\frac{a}{r_1} \sin(kr_1) + \frac{a}{r_2} \sin(kr_2+\phi)\right]^2. \qquad (22\text{-}6)$$

It is usually worthwhile to go one more step and also subtract kr_1 from both angles.
Then

$$E_0{}^2 = \left\{\frac{a}{r_1} + \frac{a}{r_2} \cos\left[k(r_2-r_1)+\phi\right]\right\}^2 + \left\{\frac{a}{r_2} \sin\left[k(r_2-r_1)+\phi\right]\right\}^2. \qquad (22\text{-}7)$$

This form highlights the importance of the difference in the phases of the waves at the observation point. The phase difference is $(kr_2-\omega t+\phi)-(kr_1-\omega t)=k(r_2-r_1)+\phi$.

In most situations, r_1 and r_2 are both large compared to the wavelength, but their difference is not much different from λ. Significance is sometimes lost in the calculation if $k(r_2-r_1)$ is calculated by subtracting r_1 from r_2. It is sometimes possible, as we shall see, to compute r_2-r_1 directly without subtracting two large numbers. When this is possible, Eqn. 22-7 is preferable to Eqn. 22-6.

For all of the situations considered here, the sources are of equal strength. Since we are interested in the variation of the intensity as the observation point moves from place to place, the strength of the sources is immaterial. To reduce the complexity of the program used, we take a to be unity.

Once E is put in the form $E=E_0\sin(-\omega t+\alpha)$, it is easy to find the average of E^2 over one cycle. The period is $2\pi/\omega$ and the average of E^2 is

$$(E^2)_{ave} = \frac{\omega}{2\pi} \int_0^{2\pi/\omega} E_0{}^2 \sin^2(-\omega t+\alpha)\, dt = \tfrac{1}{2}E_0{}^2. \qquad (22\text{-}8)$$

Since we are interested in how the intensity changes from point to point and not in the absolute magnitude of the intensity, we omit the factor $\tfrac{1}{2}$ and calculate $E_0{}^2$. For the problems of this chapter, we plot $E_0{}^2$ as a function of the position of P on a viewing screen at x=D.

Given D, d, and Δ, r_1 is found using

$$r_1 = \left[D^2 + (d-\Delta)^2 \right]^{\frac{1}{2}}, \tag{22-9}$$

and r_2 is found using

$$r_2 = \left[D^2 + (d+\Delta)^2 \right]^{\frac{1}{2}}. \tag{22-10}$$

Then $E_0{}^2$ is calculated using Eqn. 22-7.

When using the program of Fig. 2-4 (Appendix C) to plot $E_0{}^2$, the independent variable is d and its initial value is entered into X1. Its final value is entered into X2 and the increment is entered into X3. The two parameters are D and Δ and their values are entered into X4 and X5, respectively.

To use Eqn. 22-7, line 120 is replaced by a series of steps. First r_1 and r_2 are computed and stored in X7 and X8 respectively, then the phase difference is computed and stored in X9. The instructions are

SQRT(X4↑2+(X1−X5)↑2)→X7

SQRT(X4↑2+(X1+X5)↑2)→X8

and   (2π/λ)*(X8−X7)+(φ)→X9

Next, the quantities in the brackets of Eqn. 22-7 are evaluated, using

(1/X7)+(1/X8)*COS(X9)→X7

and   (1/X8)*SIN(X9)→X8

These quantities are then squared and added to produce $E_0{}^2$:

X7↑2+X8↑2→X6

Values for λ and ϕ must be substituted into these instructions when the machine is programmed.

The program is now used to examine some interference patterns.

When kr_1 and $kr_2+\phi$ differ by an integer multiple of 2π radians, $\cos(kr_1)$ and $\cos(kr_2+\phi)$ have the same value, as do $\sin(kr_1)$ and $\sin(kr_2+\phi)$, and the intensity is nearly a maximum. When kr_1 and $kr_2+\phi$ differ by an odd integer multiple of π radians, $\cos(kr_1)$ and $\cos(kr_2+\phi)$ have the same magnitude but opposite sign. The same statement can be made about $\sin(kr_1)$ and $\sin(kr_2+\phi)$, so the intensity is nearly zero. In each case, the word nearly must be used because r_1 and r_2 appear in the denominators of Eqn. 22-7.

As the observation point P moves along the viewing screen, both r_1 and r_2 change and, because the sources are separated, they change by different amounts. The difference $k(r_2-r_1)+\phi$ continually increases as P moves from the x axis up the screen and, as it increases, it attains values which are alternatively even and odd integer multiples of π. The intensity is alternatively bright and dark.

In the usual situation, P need not change much for the intensity to go from bright to dark. In fact, r_2-r_1 need change by an amount which is only half the wavelength of the light. Since visible light has wavelengths of about 5×10^{-7} m, you can easily understand why a small change in r_2-r_1 can produce a large change in the intensity.

Because the screen is removed from the sources, d must change by a larger amount than r_2-r_1 to go from the center of a bright area to the center of a dark area. As the screen is brought closer to the sources, d must change less to achieve the same result. In addition, to go from high to low intensity, d must change by different amounts for different regions of the screen. In general, at large d, the bright regions of the screen are wider than for small d.

The following problem illustrates these assertions.

<u>Problem 1.</u>   Two point sources, separated by $2\Delta=1\times10^{-4}$ m, emit spherical waves with the same amplitude and with wavelength $\lambda=5\times10^{-7}$ m, in phase with each other ($\phi=0$). The intensity pattern is observed on a screen 2 m away, oriented as shown in Fig. 22-1.

a.   Use the program of Fig. 2-4 to plot E_0^2 for values of d between 0 and 2×10^{-2} m, with an interval width of 1×10^{-3} m. Take a=1.

b.   Use the program of Fig. 2-4 to plot E_0^2 for values of d between 2 m and 2.02 m with an interval of 1×10^{-3} m. Take a=1.

c.   For each of the situations of parts a and b, calculate $k(r_2-r_1)$ when the observation point is near a maximum of intensity and when the observation point is near a minimum of intensity. Are the results as you expect?

418

d. Explain why the bright regions are so much wider near d=2 m than they are near d=0. To do this, calculate several values of r_2-r_1 for values of d differing by 1×10^{-3} m.

When the screen is brought closer to the sources, the intensity pattern is somewhat different.

Problem 2. Two point sources, of equal strength and separated by $2\Delta = 1\times10^{-4}$ m, emit spherical waves with wavelength $\lambda=5\times10^{-7}$ m, in phase with each other ($\phi=0$). The intensity pattern is observed on a screen 0.01 m away, oriented as shown in Fig. 22-1.

a. Use the program of Fig. 2-4 to plot E_0^2 for values of d between 0 and 1×10^{-4} m with an interval of 5×10^{-6} m. Take a=1.

b. Calculate $k(r_2-r_1)$ for situations when the observation point is near a maximum of the intensity and when the observation point is near a minimum of the intensity. Are the results as you expect?

The pattern has much the same shape as it has for the situation of problem 1. Now, however, it is much compressed. The maxima and minima are closer together.

Notice that, for the two problems, the division by r_1 and r_2 which occurs in Eqn. 22-7 seems to have little effect on the positions of the maxima and minima of intensity. This is because Δ is so much smaller than D and r_1 is very nearly the same as r_2. The difference in the values of r_1 and r_2 is important for the evaluation of the trigonometric functions which appear in Eqn. 22-7, however.

We now derive an approximate expression for the intensity, valid when the separation between the sources is small compared to the distance to the point of observation. Although the expression to be derived is only an approximation, it may give more accurate results than the exact expression when $D^2+d^2>>\Delta^2$. In calculating r_2-r_1, you may have noticed some loss of significance because r_1 and r_2 differ only in the 7th or 8th significant figure. The new technique allows the difference to be calculated without performing the subtraction numerically.

If the difference between r_1 and r_2 is small, great simplification can be made in the calculation of the intensity. First, Eqn. 22-9 is approximated by

$$r_1 = r - \Delta\sin\theta \qquad\qquad (22\text{-}11)$$

where r and θ are defined in Fig. 22-1. This result is obtained from Eqn. 22-9 by writing

$$r_1 = \left[(D^2+d^2) - 2d\Delta + \Delta^2\right]^{\frac{1}{2}}, \qquad\qquad (22\text{-}12)$$

an equation which results when the quantity in the parentheses is squared and the terms rearranged. The binomial theorem expansion of $(A+B)^{\frac{1}{2}}$, where A is much larger than B, is

$$(A+B)^{\frac{1}{2}} = A^{\frac{1}{2}} + \tfrac{1}{2}A^{-\frac{1}{2}}B + \dots \qquad\qquad (22\text{-}13)$$

Take $A=(D^2+d^2)$, a large number, and $B=-2d\Delta+\Delta^2$, a small number. Ignore terms which contain Δ^2 or higher powers of Δ. Then

$$r_1 = \left[(D^2+d^2) - 2d\Delta + \Delta^2\right]^{\frac{1}{2}} = (D^2+d^2)^{\frac{1}{2}} - \tfrac{1}{2}\frac{2d\Delta}{(D^2+d^2)^{\frac{1}{2}}} + \dots \qquad (22\text{-}14)$$

Now $(D^2+d^2)^{\frac{1}{2}}=r$ and $d/(D^2+d^2)^{\frac{1}{2}}=\sin\theta$, so

$$r_1 = r - \Delta\sin\theta. \qquad\qquad (22\text{-}15)$$

Similarly,

$$r_2 = r + \Delta\sin\theta. \qquad\qquad (22\text{-}16)$$

In the denominators of Eqn. 22-7, take $r_1=r_2=r$ (ignore terms which contain Δ), but retain the terms in Δ when evaluating the trigonometric functions. Use $r_2-r_1=2\Delta\sin\theta$, then Eqn. 22-7 becomes

$$E_0^{\,2} = \frac{a^2}{r^2}\left[1 + \cos(\beta+\phi)\right]^2 + \frac{a^2}{r^2}\sin^2(\beta+\phi), \qquad (22\text{-}17)$$

where $\beta=2k\Delta\sin\theta$, $r=(D^2+d^2)^{\frac{1}{2}}$, and $\sin\theta=d/r$.

This result can be simplified a little more. Square the quantity in the first set of brackets, use $\cos^2(\beta+\phi)+\sin^2(\beta+\phi)=1$, then use $\cos(\beta+\phi)=2\cos^2(\frac{\beta+\phi}{2}) - 1$ to write

$$E_0^{\,2} = 4\,\frac{a^2}{r^2}\cos^2(\frac{\beta+\phi}{2}). \qquad\qquad (22\text{-}18)$$

420

To check the validity of this expression, we repeat problem 1.

Problem 3.   Consider the source and screen of problem 1 and use the program of Fig. 2-4 to plot E_0^2 for the same points as were used in part a of that problem. This time, however, use Eqn. 22-18 to evaluate E_0^2. For each value of d, have the machine compute r and $\sin\theta$, then E_0^2. Take a=1. Compare the results with those of problem 1.

If the phase angle ϕ is not zero, the intensity pattern shifts relative to the pattern when it is zero. The maximum which was at a point such that $\beta=0$ is now at a point such that $(\beta+\phi)=0$.

Problem 4.   Two point sources, separated by $2\Delta=1\times10^{-4}$ m, emit spherical waves with the same amplitude and with wavelength $\lambda=5\times10^{-7}$ m, out of phase with each other. Take $\phi=1.2$ radians in Eqn. 22-18. The intensity pattern is viewed on a screen 2 m away, oriented as shown in Fig. 22-1.
a.   Use the program of Fig. 2-4 to plot E_0^2 for values of d between 0 and 2×10^{-2} m, with an interval width of 1×10^{-3} m. Take a=1.
b.   Since $1.2/2\pi\approx0.19$, the position of a maximum should be shifted about one fifth the distance between successive maxima. The shift is downward on the screen of Fig. 22-1. Compare the results of part a with those of problem 1 and verify that this has indeed occurred. The proportionality used here is not exactly true but it is accurate as long as r_1 and r_2 are nearly the same.

Eqn. 22-7 can easily be generalized for the case of more than two sources. Suppose there are N sources, all emitting spherical waves with the same amplitude and wavelength. Then, if the observation point is a distance r_i from source i, the disturbance due to that source is given by

$$\frac{a}{r_i}\sin(kr_i-\omega t+\phi_i).\qquad(22\text{-}19)$$

The total disturbance at the observation point is

$$E=\sum_{i=1}^{N}\frac{a}{r_i}\sin(kr_i-\omega t+\phi_i).\qquad(22\text{-}20)$$

This sum has the form $E_0\sin(-\omega t+\alpha)$, where

$$E_0^2 = \left[\sum_{i=1}^{N} \frac{a}{r_i} \cos(kr_i+\phi_i)\right]^2 + \left[\sum_{i=1}^{N} \frac{a}{r_i} \sin(kr_i+\phi_i)\right]^2 . \qquad (22\text{-}21)$$

You should verify that this result reduces to Eqn. 22-7 for the case of two sources.

If source i is located at the point x_i,y_i in the x,y plane, the viewing screen is the plane x=D, and the observation point is a distance d from the x axis, then

$$r_i = \left[(D-x_i)^2 + (d-y_i)^2\right]^{\frac{1}{2}}. \qquad (22\text{-}22)$$

Furthermore, if the sources are all near the origin and $D^2+d^2 >> x_i^2+y_i^2$, then

$$r_i \simeq r - (x_i\cos\theta + y_i\sin\theta), \qquad (22\text{-}23)$$

where $r=(D^2+d^2)^{\frac{1}{2}}$ is the distance from the origin to the observation point and θ is the angle between the x axis and the line from the origin to the observation point. The trigonometric functions can be evaluated using $\cos\theta=D/r$ and $\sin\theta=d/r$. For this situation it is again valid to take $r_i=r$ for all the sources when evaluating the denominators of Eqn. 22-21. The arguments of the trigonometric functions are given by $kr-k(x_i\cos\theta + y_i\sin\theta) + \phi_i$. They contain the common term kr which, if omitted, changes the phases of all the waves by the same amount but does not change the amplitude of the resultant. For situations in which $D^2+d^2 >> x_i^2+y_i^2$,

$$E_0^2 = \frac{a^2}{r^2}\left[\sum_{i=1}^{N}\cos\beta_i\right]^2 + \frac{a^2}{r^2}\left[\sum_{i=1}^{N}\sin\beta_i\right]^2, \qquad (22\text{-}24)$$

where $\beta_i=k(x_i\cos\theta + y_i\sin\theta) + \phi_i$.

When the observation point is far away from the sources, it is usual to multiply Eqn. 22-24 by r^2 and consider $r^2E_0^2$, rather than E_0^2. The right side of the equation then depends on θ and not on r. This function is used to calculate the relative intensity for points nearly the same distance from the origin.

Problem 5. Three identical sources emit spherical waves with wavelength $\lambda=5.5\times10^{-7}$ m and all of the waves are in phase at the sources. The sources are situated on the y axis at $y=-1.2\times10^{-4}$ m, 0, and $+2.4\times10^{-4}$ m, respectively. The viewing screen is far away.

a.   Plot $r^2E_0^2$ as a function of the angle θ. Use points every 2.5×10^{-4} radians from $\theta=0$ to $\theta=5\times10^{-3}$ radians. To use the program of Fig. 2-4, place the initial value of θ in X1, the final value in X2, and the interval width in X3. The two parameters, in X4 and X5, are not used. Take $a=1$.

b.   For each of the following values of θ, calculate the phase, far from the sources, of the waves from the second and third sources, relative to the phase of the wave from the first source. These are given by $1.2\times10^{-4}k\sin\theta$ and $3.6\times10^{-4}k\sin\theta$ respectively. Here k is in reciprocal meters.

   i. $\theta=0$.

   ii. $\theta=1.6\times10^{-3}$ radians.

   iii. $\theta=3\times10^{-3}$ radians.

   iv. $\theta=4.6\times10^{-3}$ radians.

   v. $\theta=1\times10^{-3}$ radians.

c.   For each of the values of θ given in part b, draw a phasor diagram showing three phasors of equal magnitude, with the first phasor along the x axis and each of the others in appropriate directions as determined by the relative phases of the waves. In particular, the second phasor makes the angle $\beta_2-\beta_1$ with the x axis and the third phasor makes the angle $\beta_3-\beta_1$ with the x axis. In each case, draw the resultant.

d.   Use the phasor diagrams to explain why two of the maxima found are larger than the two maxima between them. The values of θ given are near the correct values for maxima or minima.

It is important that the phases of the component waves remain constant over periods of time which are long enough for observation of the intensity pattern. Two light bulbs, arranged as were the sources of problem 1, do not produce an observable interference pattern. In front of the light bulbs, the intensity is roughly twice the intensity of either of the bulbs.

To give a simple example, suppose that the point sources of problem 1 each emit light in short bursts. Suppose further that each burst is a sinusoidal wave with a definite phase constant but different bursts from the same source have different phase constants. At any instant of time, the two waves reaching the observer create an interference pattern. There are regions where the waves are in phase and the intensity is nearly four times the intensity of one source. There are regions where the waves are 180° out of phase and the intensity is nearly zero.

A short time later, however, the phases of the waves are both different. When the next set of bursts reaches the observer, there is another interference pattern with differently positioned bright and dark regions. As time goes on, the interference pattern changes as different sets of bursts reach the observer. If these changes occur at sufficiently short time intervals, the observer sees the average intensity and the interference pattern is washed out. The waves are said to be incoherent.

At any point, the average of the shifting intensity pattern is very nearly the sum of the intensities produced by the individual sources. The more bursts that contribute to the average, the closer the intensity becomes to this sum.

The following problem is designed to show you how the interference pattern becomes washed out. To work the problem you will need a source of numbers which are chosen randomly from the set between 0 and 2π. Many machines have the capability to generate random numbers, usually between 0 and 1000. If yours does not, use the list given in Appendix A. When a random number is called for, select one, divide it by 1000 and multiply the result by 2π. This places the result within the proper range. When another random number is needed, take it in order as produced by the machine or in order from the list.

Problem 6.   Consider the two point sources of problem 1, but choose the phase constant by randomly selecting a number between 0 and 2π. Add X10 to the list in the ENTER statement at line 100 of the program of Fig. 2-4 and, when running the program, enter the phase constant into that location. The instructions for line 120 are given just prior to problem 1. In the third instruction, X10 should be substituted for ϕ.

a.   Calculate E_0^2 for d=0. The intensity is no longer four times as strong as the intensity due to a single source, nor is it twice as strong. Interference effects do occur but there is no longer a strong maximum at d=0. It is elsewhere.

b.   Now suppose the phase constant suddenly changes. Repeat the calculation of part a, using a new, randomly selected, phase constant. Repeat again until you have 15 values for the intensity at d=0, then find the average by summing them and dividing the sum by 15. The result should be close to twice the intensity due to one of the sources. As more and more contributions are included in the average, the result becomes closer and closer to the sum of the intensities of the sources.

c.   The same phenomenon occurs at all values of d. Repeat the calculations of parts a and b for $d=5\times10^{-3}$ m, a minimum of the intensity for $\phi=0$.

22.2 Diffraction

When light passes the edge of an opaque object, part of the region in the geometric shadow is illuminated. The geometric shadow is bounded by the straight line ray from the source to the object's edge and continued beyond. It is clear that ray optics cannot account for the appearance of light in the shadow area and we must try to obtain understanding of the phenomenon from wave optics.

Interference effects can also be seen near the boundary of the geometric shadow. If the light retains the same phase constant over periods of time which are long enough for observation, a series of interference fringes, consisting of alternating bright and dark regions, is seen. A collection of waves, not just one, has reached the shadow area. The waves have different phases relative to each other and the phases are different at different places.

The diffraction effects considered here are understood in terms of Huygens' principle, which describes both the spreading of the light into the shadow region and the origin of the interference effects.

According to Huygens' principle, we may consider each point on a wave front to be the source of a spherical wave. The amplitude of the spherical wave is related to the amplitude of the original wave and the phase of the spherical wave is related to the phase of the original wave. These spherical waves are called Huygens wavelets.

At any instant of time, the original wave may be replaced by the set of Huygens wavelets and, from then on, the disturbance at any point can be found by summing the wavelets which have reached that point.

If the original wave does not meet an obstruction, there is no advantage to thinking about its propagation in terms of Huygens wavelets. The wavelets sum to form a wave which is identical to the original wave.

If there is an obstruction, however, it is useful to think about the wavelets. Consider, for example, a plane wave incident on a slit in an otherwise opaque shield and picture the situation just as a wave front gets to the slit. It takes wavelets from all points on the wave front to construct a plane wave on the other side of the shield. But the shield blocks all wavelets except those which originate at parts of the wave front at the slit. The wave on the other side of the shield is the superposition of those wavelets which are not obstructed and these do not form a plane wave.

The Huygens wavelet picture clearly explains the illumination of the geometric
shadow region. The spherical wavelets which start at points within the slit spread out
in all directions, moving along rays which are radially outward from points in the slit.
Some of these rays pass into the geometric shadow.

The wavelet picture also explains the appearance of interference fringes. The
various wavelets which simultaneously get to any selected point on the far side of the
shield started at different points in the slit and get to the selected point with
different phases. Furthermore, at different observation points, the phase relationships
are different. In some regions the wavelets add constructively to form bright fringes,
while, in other regions, they add destructively to form dark fringes.

It should be mentioned that Huygens' principle is an approximation. An exact
expression for the wave everywhere can be derived from first principles, Maxwell's
equations in the case of electromagnetic radiation. It is a rather complicated
expression, but it reduces to the expression found using Huygens' principle in all cases
we consider. The exact expression must be used when the surface considered, the slit in
the above example, is close to the light source and close to the observation point. The
greater these distances are, the more exact is Huygens' principle. Usually a few
hundred wavelengths is a large enough distance to obtain sufficient accuracy with the
approximation.

Now to the mathematical details.

Consider plane waves incident on a slit which is oriented perpendicular to the
rays as shown in Fig. 22-3. At any instant of time, part of a wave front is within the
slit and we concentrate on the spherical wavelets which emanate from that part of the
wave front. The slit lies along the y axis and y´ is the coordinate of a point in the
slit. The observation point has the coordinates x,y.

According to Huygens' principle, the wavelet from y´ has the form

$$E = \frac{B}{r} (1 + \cos\theta) \sin(kr - \omega t), \qquad (22\text{-}25)$$

where r is the distance from the wavelet source to the observation point and θ is the
angle between the ray from the source to the observation point and the x axis. B is a

426

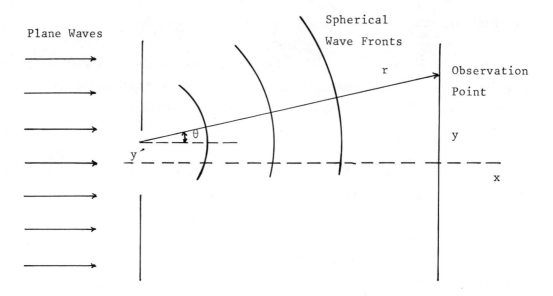

Figure 22-3.   Plane waves incident on a single slit. Some wave fronts for one
spherical Huygens wavelet are shown along with a ray from the source
of the wavelet to the observation point. Spherical wavelets emanate
from all points within the slit.

constant and is chosen so that, in the absence of an obstacle, the wavelets sum to
produce a plane wave with the same amplitude as the original plane wave. Since we are
not concerned with the absolute intensity, but only with changes in the intensity as the
observation point moves, we shall choose its value for computational convenience.

The amplitude of the spherical wavelet is not the same as the amplitude of the
incident plane wave. Since there are an infinite number of wavelets with sources
within the slit, B must be infinitesimal. In addition, the amplitude contains the
factor $(1+\cos\theta)$. The wavelet has a larger amplitude in the forward direction (θ near 0)
than in the backward direction (θ near 180°). This factor is important if the sum of the
wavelets is to reproduce a wave which is traveling in the same direction as the
original wave. It is also important, as we shall see, if Huygens' principle is to
produce the correct intensity pattern in the region close to the edge of an opaque
object. For many problems, we shall consider intensity patterns far from the object
and near the x axis. Then $\cos\theta \approx 1$ for all wavelets which reach the observation point
and this factor does not play an important role in determining the intensity pattern.

In principle, the wave form given by Eqn. 22-25 is incorrect in one respect.

Although the frequency and wavelength are the same as those of the original wave, there is a change in the phase constant. The phase of the wavelet differs from that of the original wave by $\pi/2$. This phase change does influence the phase of the disturbance at the observation point, but since it is the same for all of the wavelets, it does not influence the intensity pattern there. It is required, however, if the sum of the wavelets is to reproduce the original wave. We have omitted it since we are interested only in the intensity pattern.

The wavelet is spherical and the amplitude decreases as $1/r$. Since the coordinates of the observation point are x,y,

$$r = \left[x^2 + (y-y')^2 \right]^{\frac{1}{2}} \tag{22-26}$$

and

$$\cos\theta = x/r. \tag{22-27}$$

All of the wavelets which originate at points in the slit must now be summed to find the total disturbance at the observation point. Since there is a continuous distribution of wavelet sources in the slit, the sum must take the form of an integral. The amplitude of a wavelet must be infinitesimal and we take $B=dy'/a$ where a is the width of the slit.

The total disturbance at the observation point is then given by the integral

$$E = (1/a) \int_{-a/2}^{+a/2} (1/r)(1+\cos\theta) \sin(kr-\omega t) \, dy'. \tag{22-28}$$

Remember that θ and r are functions of y'. The resultant wave has the form $E=E_0\sin(-\omega t+\alpha)$ and its amplitude can be found in a manner similar to that used to derive Eqn. 22-21. Sums which appear in that equation are now written as integrals and the result is

$$E_0^2 = (1/a)^2 \left[\int_{-a/2}^{+a/2} (1/r)(1+\cos\theta) \sin(kr) \, dy' \right]^2$$

$$+ (1/a)^2 \left[\int_{-a/2}^{+a/2} (1/r)(1+\cos\theta) \cos(kr) \, dy' \right]^2. \tag{22-29}$$

The intensity is proportional to E_0^2.

428

The integration program of Fig. 8-1 (Appendix C) can be used to evaluate the integrals which appear in Eqn. 22-29. It must be run twice, once for each integral. The variable of integration is y'. The lower limit $-a/2$ is entered into X1 and the upper limit $+a/2$ is entered into X2. Since the program will be run many times, with different values of y, it is useful to store y in X10, for example, and recall it as needed rather than include its value as part of the program instructions. X10 is added to the list of the ENTER statement at line 100. At line 130, the appropriate instructions for evaluation of the first integral are

$$\text{SQRT}(x{\uparrow}2+(X10-X1){\uparrow}2){\to}X5$$

$$(1+x/X5)*\text{SIN}((2\pi/\lambda)*X5)/X5{\to}X5$$

where values for x and λ must be inserted when the machine is programmed. The instructions at lines 170 and 180 are the same, except X5 is replaced by X9. For the second integral, COS is substituted for SIN in the second line above. Once both integrals have been evaluated, the results are squared and summed, then the sum is multiplied by $(1/a)^2$.

A great saving in running time is achieved if the two integrals are evaluated together. Then kr and $(1+\cos\theta)/r$ need be calculated only once for each of the integration intervals. This requires a larger machine memory, however. To evaluate both integrals within the same program, the following memory allocation can be made:

X1: y' (starting value: $-a/2$),
X2: $a/2$,
X3: N,
X4: $\Delta y'$,
X5: $(1/r)(1+\cos\theta) \sin(kr)$ for $y'=0$,
X6: collects odd terms for integral with sine function,
X7: collects even terms for integral with sine function,
X8: counter,
X9: stores various intermediate results,
X10: $(1/r)(1+\cos\theta) \cos(kr)$ for $y'=0$,
X11: collects odd terms for integral with cosine function,
X12: collects even terms for integral with cosine function,
X13: stores various intermediate results,
X14: r, then $(1+\cos\theta)/r$,

X15: kr,

and   X16: y.

At line 130, the following steps can be used to compute the contributions to the two integrals:

$$\text{SQRT}(x{\uparrow}2+(X16-X1){\uparrow}2) \rightarrow X14$$
$$X14*(2\pi/\lambda) \rightarrow X15$$
$$(1+x/X14)/X14 \rightarrow X14$$
$$\text{SIN}(X15)*X14 \rightarrow X9$$
$$\text{COS}(X15)*X14 \rightarrow X13$$
$$X9 \rightarrow X5$$
$$X13 \rightarrow X10$$

First r is calculated, then kr and $(1+\cos\theta)/r$. In the next step, $\sin(kr)$ is multiplied by $(1+\cos\theta)/r$ and the result is placed in X9. Then $\cos(kr)$ is multiplied by $(1+\cos\theta)/r$ and the result is placed in X13. X9 and X13 now contain the contributions of the interval to the two integrals and they are stored for later use, in X5 and X10 respectively. Values for x and λ must be supplied when the machine is programmed.

At line 170, the instructions are the same, except that the last two are replaced by

$$X6+X9 \rightarrow X6$$
and
$$X11+X13 \rightarrow X11$$

respectively. Again, at line 180, the same instructions are used, except that the last two are replaced by

$$X7+X9 \rightarrow X7$$
and
$$X12+X13 \rightarrow X12$$

respectively. X11 and X12 must be initialized to zero before the loop is entered, at line 140. In addition, the value for y must be entered into X16 at line 100.

When the loop has been traversed N times, the various contributions to the second integral must be combined. After the instruction already there, add

$$(X10+4*X11+2*X12-X13)*(X4/3) \rightarrow X9$$

to line 200. The contributions to the first integral are combined by means of instructions in the program as it stands and the result is stored in X8. It is also convenient to add the squares of the two integrals. Add the instruction

$$X8\uparrow 2+X9\uparrow 2 \rightarrow X10$$

to line 200. This value is then displayed. It should be multiplied by $(1/a)^2$, either before or after the result is displayed.

Problem 1.   A 1×10^{-4} m wide slit is illuminated by plane waves with wavelength 5×10^{-7} m and the intensity pattern is viewed on a screen 1 m away. Use the numerical integration program, with N=16, to plot E_0^2 every 1×10^{-3} m from $y=-15 \times 10^{-3}$ m to $y=+15 \times 10^{-3}$ m. The intensity for negative y is exactly the same as that for positive y, so carry out the calculation only for $y \geq 0$. Indicate on the graph the geometric image of the slit. It extends from $y=-5 \times 10^{-5}$ m to $y=+5 \times 10^{-5}$ m.

The pattern is dominated by a broad, intense, central fringe, with a maximum at y=0. This is followed, on either side, by a series of less intense bright fringes, separated by regions where the intensity is low. The fringes, of course, come about as a result of the interference of the Huygens wavelets. At y=0, the wavelets are all nearly in phase with each other and add constructively. As y increases, the various wavelets travel different distances to the observation point and arrive with different phases. At a minimum of the intensity, the phases are such that the sum of the wavelets is small.

Notice that the intensity pattern spreads well beyond the region of the geometric image. In fact, the central bright band alone is roughly 100 times as wide as the geometric image. The appearance of secondary maxima makes the pattern even wider.

In practice, numerical integration need not be used for the above problem. When the viewing screen is far enough from the slit, as it is for that problem, certain approximations can be made in Eqn. 22-29 and, when they are made, the integrals can be evaluated in closed form.

When the distance $(x^2+y^2)^{\frac{1}{2}}$ from the center of the slit to the observation point is large compared to the slit width a, r can be approximated by $r=r_0-y'\sin\theta$, where $r_0=(x^2+y^2)^{\frac{1}{2}}$. This expression can be derived by applying the binomial theorem to Eqn. 22-26 and recognizing that $y/r_0=\sin\theta$. The steps are

$$\left[x^2+(y-y')^2\right]^{\frac{1}{2}} = \left[(x^2+y^2)-2yy'+y'^2\right]^{\frac{1}{2}}$$

$$= (x^2+y^2)^{\frac{1}{2}} + \tfrac{1}{2}(x^2+y^2)^{-\frac{1}{2}}(-2yy'+y'^2) + \dots$$

$$= r_0 - \frac{yy'}{r_0} + \dots$$

$$\simeq r_0 - y'\sin\theta. \tag{22-30}$$

Here terms which contain the square and higher powers of y' are neglected since y'/r is small.

We next assume the observation point is close to the x axis so $\cos\theta\simeq1$. Then the factor $(1+\cos\theta)/r$ is nearly $2/r$. When we substitute the approximate expression for r into $\sin(kr)$ and $\cos(kr)$, we may omit the term kr_0 since this phase angle is a constant for both integrals and does not influence the amplitude of the wave at the observation point.

When all these approximations are made, Eqn. 22-29 becomes

$$E_0^2 = (\frac{2}{ar_0})^2\left[\int_{-a/2}^{+a/2}\sin(ky'\sin\theta)\,dy'\right]^2 + (\frac{2}{ar_0})^2\left[\int_{-a/2}^{+a/2}\cos(ky'\sin\theta)\,dy'\right]^2 \tag{22-31}$$

and the integrals can be evaluated analytically. θ is constant. The first integral vanishes and

$$E_0^2 = (\frac{2}{ar_0})^2\frac{4\sin^2(k\frac{a}{2}\sin\theta)}{k^2\sin^2\theta}. \tag{22-32}$$

This is usually written

$$E_0{}^2 = B^2 \frac{\sin^2\beta}{\beta^2} \qquad (22\text{-}33)$$

where $\beta = k\frac{a}{2}\sin\theta$ and B is constant for observation points near the x axis.

The approximations made to obtain this result can be interpreted physically. The observation point is so far from the slit that the wavelets are nearly plane waves and the curvature in the spherical wave fronts can be neglected. The shape of the intensity pattern no longer depends on x and $E_0{}^2$ is a function of the angle θ alone.

According to Eqn. 22-33, $E_0{}^2 = 0$ when $\beta = n\pi$, where n is any integer but 0. Since $\beta = k\frac{a}{2}\sin\theta$, this means $\sin\theta = n\lambda/a$ and, since $\sin\theta = y/r_0$, the minima of the intensity occur at

$$y = r_0\, n\lambda/a. \qquad (22\text{-}34)$$

Notice that the condition for a zero of intensity is that the wavelets from the top and bottom of the slit travel distances which differ by an integer number of wavelengths. They arrive at the screen in phase. For one of these values of θ, any two wavelets with source points a/2 apart are 180° out of phase at the screen and sum to zero there. The wavelets can be paired in this way, with one wavelet of any pair coming from the bottom half of the slit and the other coming from the top half, a distance a/2 away. The sum of all the wavelts must be zero.

The maxima in the intensity pattern occur when β is about $n\pi/2$, where n is an odd integer. This condition causes the numerator of Eqn. 22-33 to have its largest value, but the condition is not exact because β appears in the denominator. Exact positions of the maxima can be found be differentiating Eqn. 22-33 with respect to β, setting the derivative equal to zero, and solving for β. The coordinate of a maximum on the screen is then given by $y = r_0\sin\theta = r_0\, 2\beta/(ka)$. Numerical techniques must be used to find exact values of β.

Problem 2. Consider the slit and viewing screen of problem 1. By answering the following questions you will compare Eqn. 22-33 with the intensity function obtained by numerical integration. If they compare favorably, you should be convinced that the approximations which lead to Eqn. 22-33 are valid for the slit width, slit-to-screen distance, and wavelength of this problem.

a. Use Eqn. 22-33 to find the positions on the screen of the first three zeros of the intensity pattern on one side of the central maximum. Compare your answers with the zeros of the graph you plotted in response to problem 1.

b. Show that the derivative with respect to β of Eqn. 22-33 is

$$dE_0^2/d\beta = \frac{2B^2}{\beta^3} \sin\beta \left[\beta\cos\beta - \sin\beta \right] .$$

Use the program of Fig. 2-3 (Appendix C) to find values of β for which this derivative vanishes. The function to be considered is $\beta\cos\beta - \sin\beta$ and, for a given maximum, the limits of the search are the values of β which are near minima on either side. In the search, avoid $\beta = 0$ and values of β for which the intensity is a minimum. Find values of β and the positions of the maxima on the viewing screen for the first two maxima on one side of the central maximum. Obtain three significant figure accuracy. Compare your answers with the maxima shown on the graph you plotted in response to problem 1. The maxima are not halfway between the minima, but are displaced slightly toward the center of the pattern.

c. For each value of β found in part b, substitute that value into Eqn. 22-33 and calculate the intensity at the maximum. Take B^2 to have the value of the intensity at $y=0$, found in answer to problem 1. Compare your answers with appropriate values from the graph of problem 1.

As the slit width is narrowed, the pattern spreads.

Problem 3. A 7×10^{-5} m wide slit is illuminated by plane waves with wavelength 5×10^{-7} m and the intensity pattern is viewed on a screen 1 m away.

a. Use Eqn. 22-33 to find the positions of the first three zeros of the intensity pattern on one side of $y=0$.

b.   Calculate the ratio of the width of the central bright region for $a=7\times10^{-5}$ m to the width for $a=1\times10^{-4}$ m. Calculate the ratio of the width of the first secondary bright fringe for $a=7\times10^{-5}$ m to the width for $a=1\times10^{-4}$ m.

An increase in the wavelength also produces a broadening of the intensity pattern.

Problem 4.   A 1×10^{-4} m wide slit is illuminated by plane waves with wavelength 6.5×10^{-7} m and the intensity pattern is viewed on a screen 1 m away.

a.   Use Eqn. 22-33 to find the first three zeros of the intensity pattern on one side of y=0.

b.   Calculate the ratio of the width of the central bright fringe for $\lambda=6.5\times10^{-7}$ m to the width for $\lambda=5\times10^{-7}$ m.

c.   For what value of λ are the zeros of the intensity pattern for this slit at the same places as they are for the slit, light, and screen of problem 3?

The diffraction pattern studied in problems 1 and 2 does not look like the geometric image of the slit, as we might expect if we ignored wave optics. The geometric image is a bright band as wide as the slit and positioned on the screen opposite the slit. Rather than this image, the light passing through the slit produces a series of fringes which extend well beyond the image region.

As the slit is widened or the viewing screen is brought closer to the slit, the fringe system narrows. Eventually, the central maximum approximately occupies the region of the geometric image and very little light reaches regions of the geometric shadow. There is still some fringing near the edges of the shadow region, however.

Something else happens as the slit is widened. The central area is not uniformly illuminated, but rather fringes appear there. The intensity at the minima within the image region is not zero so the fringes are not as noticeable as the fringes of a narrow slit.

For a wide slit, different wavelets travel appreciably different distances to the observation point and their rays make appreciably different angles with the x axis. The factor $(1+\cos\theta)/r$ which appears in Eqn. 22-29 is important and it is this equation, rather than Eqn. 22-33, which must be used to evaluate E_0^2.

The following problems are designed to show you how the intensity pattern changes as the slit widens. They require considerable running time on hand held machines. To shorten the running time, the number of integration intervals used has been reduced with loss of accuracy in the third significant figure. If a fast machine is available and higher accuracy is desired, double the number of integration intervals in each case.

Problem 5. Plane wave light with wavelength 5×10^{-7} m illuminates a 7×10^{-5} m wide slit and the intensity pattern is viewed on a screen 0.02 m from the slit. Calculate E_0^2 as a function of the distance y from the x axis to the observation point. Use the program of Fig. 8-1, suitably modified, and take N=16. Plot points every 2×10^{-5} m from y=0 to $y=36 \times 10^{-5}$ m. The pattern is symmetric about y=0. On the graph, locate the edge of the slit's geometric image.

Qualitatively, the pattern is quite similar to the pattern obtained in answer to previous problems. The central bright region extends a considerable distance beyond the geometric image and this region is followed by a series of fringes. The influence of the $(1+\cos\theta)/r$ factor is small and is seen only in some details. The central maximum is not quite as bright and the dark regions between secondary maxima are slightly broader than for greater slit-to-screen distances.

Problem 6. The slit of problem 5 is now widened to 14×10^{-5} m. The wavelength of the light and the slit-to-screen distance are still 5×10^{-7} m and 0.02 m respectively. Use the program of Fig. 8-1, with N=32, to plot E_0^2 every 1×10^{-5} m from y=0 to $y=22 \times 10^{-5}$ m. Again, the pattern is symmetric about y=0 so points with negative y need not be plotted. On the graph, locate the edge of the slit's geometric image.

Notice that the pattern is much more narrow than the previously plotted pattern. The central bright region is now within the geometric image and the pattern has a shoulder near the edge of the image. This shoulder is a remnant of the first minimum of the pattern for the narrow slit. It occurs in the neighborhood of the place where a $\sin\theta=\lambda$. Now, however, the wavelets do not sum to zero. To be sure, there is a value of y for which wavelets from the lower edge and the center of the slit, for example, are 180° out of phase. However, these waves have different amplitudes at the screen since the factor $(1+\cos\theta)/r$ is different for each of them. Destructive interference occurs, but it is not complete.

Problem 7. The slit of problem 5 is now widened to 28×10^{-5} m. The wavelength of the light and the slit-to-screen distance remain the same. Take N=50 and use the program of Fig. 8-1, suitably modified, to plot E_0^2 every 1×10^{-5} m from y=0 to $y=20\times10^{-5}$ m. Points with negative y need not be plotted since the pattern is symmetric about y=0. Mark the location of the edge of the geometric image on the graph.

The extent of the geometric image is now discernible in the intensity pattern. There are fringes deep within the image but the intensity does not become zero at any place in that region. There is a gray central area where the intensity is about half that at the maximum but this merges into a bright region and then, as the edge of the geometric image is approached, the intensity falls off rapidly. There is fringing in the neighborhood of the image edge but the illuminated region does not extend very far beyond the edge.

If the slit width is increased more, the pattern more closely resembles the geometric image. The number of fringes in the image region increases but the variation in intensity from point to point becomes less dramatic. The sum of the wavelets in the central region becomes more plane wave like. The intensity falls off rapidly near the edge of the geometric image and, while there is still fringing, the fringes are faint and do not extend far into the shadow area.

22.3   Multiple Slits

Consider plane waves incident on two slits in an otherwise opaque barrier. The geometry is shown in Fig. 22-4. Each of the slits has width a and the distance between the centers of the slits is b. The bottom slit extends from y=-(b+a)/2

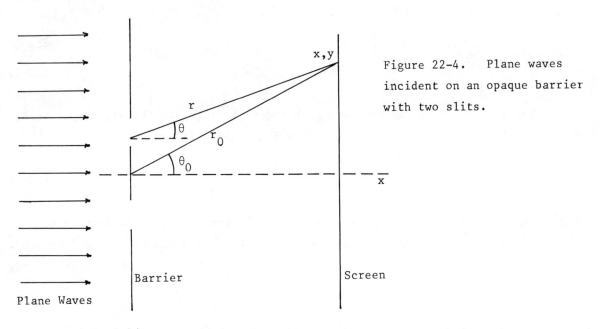

x,y

Figure 22-4.   Plane waves incident on an opaque barrier with two slits.

Barrier

Screen

Plane Waves

to y=-(b-a)/2 and the top slit extends from y=(b-a)/2 to y=(b+a)/2. The observation point is on a viewing screen behind the barrier and has coordinates x,y. The distance from a point within one of the slits to the observation point is denoted by r and the distance from the center of the slit system to the observation point is denoted by r_0.

An equation similar to Eqn. 22-29 holds, but now wavelets from both slits must be included. The intensity at the observation point is proportional to

$$E_0{}^2 = (1/a)^2 \left[\int_{-(b+a)/2}^{-(b-a)/2} (1/r)(1+\cos\theta)\sin(kr)\ dy' + \int_{(b-a)/2}^{(b+a)/2} (1/r)(1+\cos\theta)\sin(kr)\ dy' \right]^2$$

$$+ (1/a)^2 \left[\int_{-(b+a)/2}^{-(b-a)/2} (1/r)(1+\cos\theta)\cos(kr)\ dy' + \int_{(b-a)/2}^{(b+a)/2} (1/r)(1+\cos\theta)\cos(kr)\ dy' \right]^2 . \qquad (22-35)$$

Here y´ is the coordinate of a point within one of the slits. The first integral in each bracket sums the contributions of the wavelets from the bottom slit and the second integral sums the contributions of the wavelets from the upper slit.

In general, r and $\cos\theta$ are computed using

$$\dot{r} = \left[x^2 + (y-y´)^2\right]^{\frac{1}{2}} \tag{22-36}$$

and
$$\cos\theta = x/r \tag{22-37}$$

respectively. If r is large compared with the slit width and the slit separation, then an analysis similar to that preceeding Eqn. 22-30 yields

$$r \simeq r_0 - y´\sin\theta \tag{22-38}$$

where $\sin\theta = y/r_0$ and $r_0 = (x^2+y^2)^{\frac{1}{2}}$. We may then take $r = r_0$ in the denominators of Eqn. 22-35 and omit the term kr_0 from the arguments of the trigonometric functions. It is the same for all wavelets and does not influence the intensity. Furthermore, if y´ is small, $\cos\theta \simeq 1$.

Now the integrals can be evaluated in closed form and the result is

$$E_0{}^2 = (\frac{2}{ar_0})^2 \cos^2(k\,\frac{b}{2}\,\sin\theta)\,\frac{\sin^2(k\,\frac{a}{2}\,\sin\theta)}{k^2\,\sin^2\theta} \ . \tag{22-39}$$

This expression can be written

$$E_0{}^2 = (B/r_0)^2 \cos^2\gamma\,\frac{\sin^2\beta}{\beta^2} \tag{22-40}$$

where $\gamma = k\,\frac{b}{2}\,\sin\theta$, $\beta = k\,\frac{a}{2}\,\sin\theta$, and B is a constant.

Except for the unimportant factor in the parentheses, this expression is the product of the intensity for a single slit and the intensity for two sources, separated by b. Since b must be larger than a, the $\cos^2\gamma$ factor varies more rapidly with θ than does the $\sin^2\beta/\beta^2$ factor and the pattern looks like the two source

interference pattern squeezed into the single slit diffraction pattern. The
diffraction pattern forms an envelope for the interference pattern.

Problem 1.   Two slits, each 1×10^{-4} m wide, are illuminated by plane waves with
wavelength 5×10^{-7} m. The distance between the centers of the slits is 2.7×10^{-4} m.
The intensity pattern is viewed on a screen 2 m away.

a.   Use the program of Fig. 2-4 (Appendix C) to plot $\cos^2 \gamma \, \sin^2 \beta / \beta^2$ as a function
of position on the screen. Plot points every 2×10^{-4} m from y=0 to $y = 1 \times 10^{-2}$ m.
The function is 1 at y=0. Do not use the machine to evaluate the intensity
for that point. It will have trouble since $\beta = 0$. The pattern is symmetric
about y=0.

b.   Analytically find the values of y for which $\cos^2 \gamma = 0$ and, separately, the
values of y for which $\sin^2 \beta / \beta^2 = 0$. Find only those which are less than or
equal to 1×10^{-2} m. For the minima shown on the graph of part a, compare the
values of y with those found analytically.

The graph shows the central bright region of the single slit diffraction
pattern. Clearly seen are three peaks of the double source interference pattern.
There is another peak, but it is close to the minimum of the diffraction pattern
and is not distinct on the graph. To convince yourself it is there, examine the
numerical values produced by the machine.

Except at y=0, the maxima of the intensity pattern are not at the maxima
of $\cos^2 \gamma$, but they can be found numerically by means of a root finding program.

Problem 2.

a.   Use the chain rule for differentiation and differentiate Eqn. 22-40 with
respect to θ. Show that maxima or minima occur when

$$\sin\beta \, \cos\gamma \, \cos\theta \left[a\beta \, \cos\beta \, \cos\gamma - a \, \sin\beta \, \cos\gamma - b\beta \, \sin\beta \, \sin\gamma \right] = 0.$$

One or the other of the first two factors is zero at a minimum of the intensity
and can be disregarded in a search for the maxima. Likewise, the $\cos\theta$ factor

can be disregarded.

b.   For the situation described in problem 1, use the binary search program of
     Fig. 2-3 (Appendix C) to find the three smallest positive values of θ,
     exclusive of $\theta=0$, for which the intensity is a maximum. In each case, values
     of θ near adjacent minima can be used as limits of the search. Also calculate
     the position of the maxima on the viewing screen and compare the results with
     those shown on the graph of problem 1.

c.   Evaluate $\cos^2\gamma \sin^2\beta/\beta^2$ for the values of θ found in part b. Multiply by the
     value at y=0 shown on the graph of problem 1 and compare the results with the
     intensities shown on that graph.

As the distance between the slits is narrowed, the interference pattern
spreads and fewer peaks are found inside the central bright region of the single
slit diffraction pattern.

Problem 3.   Plane waves of light with wavelength 5×10^{-7} m illuminate two slits,
each with width 1×10^{-4} m. The intensity pattern is viewed on a screen 2 m away.

a.   Find the slit separation so that the third peak of the interference pattern,
     counting the one at y=0, falls at the position of the first zero of the
     single slit diffraction pattern. Do this analytically.

b.   Use the program of Fig. 2-4 (Appendix C) to plot $\cos^2\gamma \sin^2\beta/\beta^2$ for the slit
     separation found in part a. Plot points every 5×10^{-4} m from y=0 to y=1×10^{-2} m.

These intensity patterns show no evidence of the geometric images of the
slits. Wavelets from both slits interfere at all points on the viewing screen. As
the screen is moved toward the slits, however, the pattern changes until, at
sufficiently close distances, the geometric images become evident.

The single slit patterns plotted in the last section are a clue. There,
changes in the pattern were observed as the width of the slit increased. However,
the same patterns emerge as the screen is moved inward. Eventually, the high
intensity region is wholly within the geometric image and, except for some fringing
near the edges, very little light reaches into the geometric shadow. When a second

slit is present, it is only this light which combines with similar light from the second slit.

Mathematically, the key to the change in the pattern is the $(1+\cos\theta)/r$ factor, which appears in Eqn. 22-35. Place the viewing point opposite the lower slit and consider a wavelet ray from a point in the upper slit. Not only does the wavelet travel much farther than a wavelet from the lower slit but the ray makes a steeper angle with the x axis. Both of these conditions make the amplitude of the wavelet small compared to that of a wavelet from the lower slit. The pattern opposite either slit is dominated by wavelets from that slit.

To calculate the intensity when the slit-to-screen distance is small, Eqn. 22-35 must be used. The integration program of Fig. 8-1 is run four times, once for each of the integrals or the program, as modified in the last section, is run twice, once for each of the slits. Do not square the integrals until the contributions of both slits have been summed.

<u>Problem 4.</u> Two slits, as shown in Fig. 22-4, are illuminated by plane waves with wavelength 5×10^{-7} m. The slits are 1×10^{-4} m wide and are separated by 1.7×10^{-4} m ($b=2.7\times10^{-4}$ m). Find the intensity pattern on a screen 0.01 m behind the slits. Take N=30 and plot points every 1×10^{-5} m from $y=-2\times10^{-4}$ m to $y=+2\times10^{-4}$ m. Since the pattern is symmetric, only the intensity for positive y need be calculated. On the graph, mark the edges of the geometric images of the slits.

The intensity is now small in the region between the slits but increases by a factor of more than ten toward the centers of the geometric images. It then falls rapidly at the outer edges of the images. There is fringing between the images and near their outer edges. Wavelets from both slits contribute to the total disturbance between the images and interference effects are noticeable there. The geometric images, however, are quite distinct. As the viewing screen is brought still closer, fringing effects are even less noticeable in the region between the images and the pattern opposite either slit becomes more like that of a single slit viewed from the same distance.

Chapter 23
RELATIVISTIC KINEMATICS

The Lorentz transformation equations tell how the coordinates and time of
an event, as measured in one reference frame, are related to the coodinates and time
of the same event, as measured in a second frame, moving with respect to the first.
Matrix multiplication techniques are used to carry out the transformation and to
explore some of the interesting consequences of special relativity. The concepts
presented here are important for much of the modern physics to be studied in later
chapters. Relativity theory can be found in the supplementary topics section of
both PHYSICS and FUNDAMENTALS OF PHYSICS, in Chapter II of INTRODUCTION TO SPECIAL
RELATIVITY, and in Chapter 2 of RELATIVITY AND EARLY QUANTUM THEORY.

23.1   Special Relativity

Fig. 23-1 shows two inertial frames, labelled S and S´, respectively.
The frames are oriented so that their x axes coincide and their y axes are parallel.
The z axes are also parallel and are perpendicular to the page. S´ moves with
velocity v along the x axis of S and clocks are set so that t=0 when the origins
coincide. At the time t for which the diagram is drawn, the origin of S´ has
moved a distance vt from the origin of S.

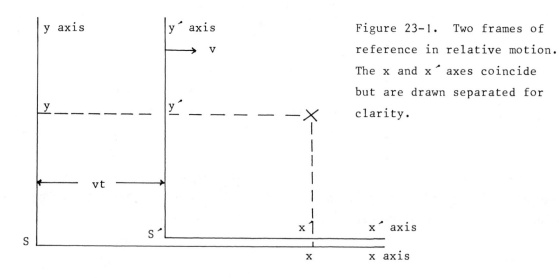

Figure 23-1. Two frames of
reference in relative motion.
The x and x´ axes coincide
but are drawn separated for
clarity.

Both S and S´ are inertial frames. That is, if zero net force acts on a particle, then its acceleration is zero, as measured in either frame. S and S´ must move with constant velocity relative to each other. If this were not so, a particle with zero acceleration, as measured in one of them, would have a non-vanishing acceleration, as measured in the other.

The place where some event occurs is marked on the diagram with an X. In S it has coordinates x,y and in S´ it has coordinates x´,y´. To specify an event, the time of occurence must be given along with the coordinates of the place where it occurs and we suppose there is a set of clocks at rest in S and another set of clocks at rest in S´ to measure time. We suppose that the event marked on the diagram occurs at time t, as measured by the clocks in S and at time t´, as measured by the clocks in S´.

It should be emphasized that the meter sticks used in S´ are exactly like those used in S and, likewise, the clocks in the two frames are similar. The instruments used in S´ are at rest relative to S´ and the instruments used in S are at rest relative to S. All the clocks in S are synchronized so that they all read the same time when any given event occurs. Similarly, all the clocks in S´ are synchronized. It is assumed that both sets of clocks are set to 0 when the origins of the two frames coincide.

Special relativity relates measurements taken with instruments at rest in S´ to those taken with instruments at rest in S. The measurements are of the same physical event. For the event shown in Fig. 23-1, relativity relates x´,y´,z´,t´ and x,y,z,t.

If the speed of S´ is slow relative to S, then the transformation equations which relate coordinate and time measurements are

$$x´ = x - vt, \qquad (23-1)$$
$$y´ = y, \qquad (23-2)$$
$$z´ = z, \qquad (23-3)$$
and
$$t´ = t. \qquad (23-4)$$

These are the transformation equations of what is called Galilean relativity.

Think of frame S as being attached to a train station and frame S´ as being attached to a moving train with a man on board. If the origin of S´ is at the back of the train, x´ gives the man's distance from the back of the train and x gives his distance from the station. If the man does not move relative to the train, x´ is constant but x increases linearly with t: x=x´+vt.

If the man walks forward with velocity u´ relative to the train, his velocity relative to the station is u=u´+v. Differentiation of Eqn. 23-1 gives this result. A man walking forward at 8 m/s in a train going 50 m/s has a speed of 58 m/s relative to the ground.

If the velocity of S´ relative to S is great, the transformation is not so simple.

The speed of light is exactly the same when measured by instruments in S´ as it is when measured by instruments in S. This, and other experimental evidence, necessitates a change in the transformation equations. If the speed of light is c in S´, then for light traveling to the right in the diagram, the Galilean equations predict it should be c+v in S. But it is not.

The correct transformation equations are known and they are called the Lorentz transformation equations. They differ greatly from the Galilean equations when v is close to the speed of light and produce nearly the same answers when v is much less than the speed of light. They will be given in the next section.

One of the most important differences between the Lorentz and Galilean transformations is that, for the Lorentz transformation, the time of an event as measured by clocks in S is not the same as the time as measured by clocks in S´. A time interval between events, as measured in S´, is different from the time interval between the same two events, as measured in S.

Fig. 23-1 depicts the situation as seen by an observer at rest in S. An observer at rest in S´ sees the frame S moving to the left with speed v (x´ component of the velocity=-v) and the origins are separated by the distance vt´. Note the prime.

The rules of kinematics, which relate coordinates and time, all measured in the same frame, are unaltered. If a particle moves with constant acceleration a along the x axis, its coordinate is given by $x(t)=x_0+u_0t+\frac{1}{2}at^2$, where x_0 is its initial coordinate and u_0 is its initial velocity. A particle may move along the x´ axis of S´ with constant acceleration a'. In that case, $x'(t')=x_0'+u_0't'+\frac{1}{2}a'(t')^2$.

The Lorentz transformation predicts some results which may seem strange to you. Two events which occur simultaneously according to clocks in one frame are not simultaneous according to clocks in another frame. When the length of a moving object is measured it is found to be shorter than when its length is measured in the frame for which it is at rest. When clocks in S are used to time ticks of a clock in S´, that clock is found to be slow but when clocks in S´ are used to time the ticks of a clock in S, it is the clock in S which is found to be slow. All these results have been verified experimentally. They will be investigated later.

Relativistic effects play a great role in the mechanics of particles, particularly at high speeds. Perhaps the most famous effect occurs when the kinetic energy of a system of particles increases by virtue of a loss in the total mass of the system. This and other relativistic phenomena are discussed in the next chapter.

23.2   The Lorentz Transformation

The situation is depicted in Fig. 23-1. Frame S´ moves with velocity v along the x axis of frame S. The origins coincide at t=0 and the clocks in S´ are set to read t´=0 then. An event occurs at x,y,z,t as measured in S, and at x´,y´,z´,t´ as measured in S´. The Lorentz transformation equations relate these measurements. The equations are

$$x' = \frac{x - vt}{(1-v^2/c^2)^{\frac{1}{2}}}, \tag{23-5}$$

$$y' = y, \tag{23-6}$$

$$z' = z, \tag{23-7}$$

and

$$t' = \frac{t - vx/c^2}{(1-v^2/c^2)^{\frac{1}{2}}} . \tag{23-8}$$

Here c is the speed of light. If frame S´ moves in the negative x direction, v is negative.

446

It is expected, of course, that x´ is not the same as x. During the time t, S´
has moved a distance vt along the x axis of S. What is perhaps not expected is that
even if the event occurs at t=0, when the origins coincide, x´ is not the same as x but
is different by the factor $1/(1-v^2/c^2)^{\frac{1}{2}}$. As we shall see, the appearance of this factor
is clearly related to the fact that a meter stick at rest in S is used to measure x while
a meter stick at rest in S´ is used to measure x´. The measurement process will be
examined in more detail later.

Distances measured along lines perpendicular to the relative velocity are the
same in both frames of reference and two of the Lorentz transformation equations are
y´=y and z´=z.

The last equation of the transformation gives the time of the event as measured
by clocks in S´. This time is not the same as the time measured by clocks in S. There
are two important differences. First t´ depends on x. This dependence describes an
important aspect of time measurements. Consider a series of events which occur at the
same time according to clocks in S but which occur at places with different x coordinates.
According to clocks in S´, these events occur at different times and the event with the
largest positive x coordinate occurs earliest.

The second difference between t and t´ is described by the appearance of the
factor $1/(1-v^2/c^2)^{\frac{1}{2}}$. The factor must be included since t is measured by clocks at rest
in S and t´ is measured by clocks at rest in S´. The time measurement process is
examined more closely later.

The Lorentz transformation equations can be written in matrix form and the
program of Fig. 21-1 can be used. In order to put the equations in more conventional
form, the abbreviations $\beta=v/c$ and $\gamma=1/(1-v^2/c^2)^{\frac{1}{2}}$ are used. Then

$$ct´ = \gamma(ct-\beta x) \tag{23-9}$$

and
$$x´ = \gamma(-\beta ct+x) \tag{23-10}$$

where the order has been changed and the equations for y´ and z´ have been omitted.

Define the two column matrices $\underline{x}'$ and $\underline{x}$ by

$$\underline{x}' = \begin{pmatrix} ct' \\ \\ x' \end{pmatrix} \quad \text{and} \quad \underline{x} = \begin{pmatrix} ct \\ \\ x \end{pmatrix} \qquad (23\text{-}11)$$

respectively, and the transformation matrix $\underline{L}$ by

$$\underline{L} = \begin{pmatrix} \gamma & -\beta\gamma \\ \\ -\beta\gamma & \gamma \end{pmatrix}. \qquad (23\text{-}12)$$

Then the transformation equations can be written

$$\underline{x}' = \underline{L} \cdot \underline{x}. \qquad (23\text{-}13)$$

To use the program of Fig. 21-1, the column matrices are expanded to 2x2 matrices by adding a column of zeros. The expanded matrix

$$\underline{x} = \begin{pmatrix} ct & 0 \\ \\ x & 0 \end{pmatrix}$$

is entered first, then $\underline{L}$. The result, calculated by the machine, is the expanded matrix

$$\underline{x}' = \begin{pmatrix} ct' & 0 \\ \\ x' & 0 \end{pmatrix}.$$

Rather than enter the elements of $\underline{L}$, it is worthwhile to reprogram the machine so that v is entered and the elements of $\underline{L}$ are computed. Store v in X9 and replace the ENTER instruction at line 120 by

```
ENTER X9, X10
X9/c→X9
1/SQRT( 1-X9↑2)→X5
X5→X8
-X9*X5→X6
X6→X7
```

The velocity v of S´ relative to S is entered and stored in X9, then β is computed. γ is calculated using $\gamma=1/(1-\beta^2)^{\frac{1}{2}}$ and then it is stored in the memory locations reserved for the first and fourth elements of $\underline{\underline{L}}$. The combination $-\beta\gamma$ is then calculated and stored in the locations reserved for the second and third elements of $\underline{\underline{L}}$. A 1 is entered in X10 if the inverse matrix is desired. Otherwise, any other number is entered. At line 150, the matrix $\underline{x}´$ is displayed, with ct´ in X1, 0 in X2, x´ in X3, and 0 in X4. To obtain t´, the number in X1 must be divided by c. When entering $\underline{x}$, it is the combination ct which is entered into X1.

The matrix $\underline{\underline{L}}$ has an inverse $\underline{\underline{L}}^{-1}$ and it is used to solve for $\underline{x}$ when $\underline{x}´$ is given. The expression is $\underline{x}=\underline{\underline{L}}^{-1}\cdot\underline{x}´$.

The following problems are designed to demonstrate some of the important features of special relativity. You should pay close attention to the results obtained. While thinking about the problems and their results, you must not think of one of the frames as being at rest and the other moving. From the point of view of S´, S is moving with speed v in the negative x´ direction. Both frames may be moving relative to a third frame.

Two events which occur at exactly the same time, as measured by clocks at rest in one frame, do not occur at the same time, as measured by clocks at rest in a second frame, moving relative to the first. The one exception is if the events occur at places with the same x coordinate. The question "are two events simultaneous?" receives a different answer from different observers, moving relative to each other.

Problem 1. At time t=0, two events occur simultaneously, according to clocks in S. Event 1 is the explosion of a firecracker at the origin and event 2 is the explosion of a firecracker at $x=5\times10^{-7}$ m. These events are viewed by an observer at rest in S´, which moves in the positive x direction at 2.6×10^8 m/s.

a. Find the coordinates and times of the two events, as measured in the primed
    reference frame. Notice that the two events do not occur simultaneously in the
    primed frame and that the distance between the events, as measured in that frame,
    is not the same as the distance as measured in the unprimed frame.

b. Flashes of light from the firecrackers spread out in all directions. Two light
    flashes, which travel along the x axis, meet at $x=2.5 \times 10^7$ m, halfway between the
    places where the events occurred. They meet at time $t=2.5 \times 10^7 /c$. Where and when
    do they meet, as measured by instruments in the primed frame?

c. In the unprimed frame, the fact that the flashes meet halfway between the sources
    is a clear indication that they started at the same time. To reach this conclusion,
    one must assume that light has the same speed no matter what its direction of
    travel. In the primed frame, the flash which started first travels the greater
    distance and the meeting place is closer to the source of the second flash. Do
    your results substantiate this conclusion? Using measurements in the primed frame
    (the results of part b), find the distance traveled by each flash from the fire-
    cracker to the place where they meet. Then divide by c to find the time taken by
    each flash. These times should differ by the same amount as the times found in
    part a.

Problem 2.   To help organize the various distance and time measurements described in
problem 1, it is worthwhile to draw diagrams of the events to show them from the points
of view of observers at rest in S and S'.

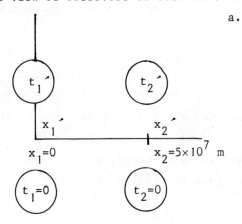

a. The diagram shows frames S and S' at time t=0, when
    the origins coincide. It also shows the events at
    $x=0$ and $x=5 \times 10^7$ m. The lower clocks are at rest
    in S and both show t=0. The upper clocks are at
    rest in S' and move to the right with speed v
    relative to S. This is the view seen by an observ-
    er at rest in S. On each S' clock, record the
    time of the event near it, as calculated in part
    a of problem 1. If either observer looks at the
    S' clocks when the events occur, these are the
    readings he sees. They are different.

Also record on the diagram the position of event 2, as measured in S'. If either
observer looks at the S' axis when the event occurs he will note that value of x' on

450

the axis opposite the event.

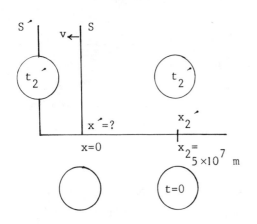

b.   The diagram shows the view from S´ when event 2 occurs. Frame S and its clocks move to the left with speed v and the clocks in S´ show the same time, the time of event 2. Fill in the values of t_2´ and the coordinate x_2´. Notice that the origins of the frames do not yet coincide but they will after a while. The event at the origin has not yet occurred but it will when the origins coincide. The clock at the origin of S does not read 0 and we wish to find its reading. To do this, first find the position of the origin of S, as measured in S´. This is just $-vt_2$´, a positive number. Now pretend there is an event at this value of x´ and at time t_2´ and use the inverse Lorentz transformation to find x and t for this event. Record the values on the diagram. Note that t is negative. You may not find that x is exactly 0 since there is round off error.

c.   Draw a diagram, similar to those drawn above, to show the situation from the viewpoint of the S´ observer when event 1 occurs. The clocks in S´ now read t_1´=0 and the clock at the origin of S reads t_1=0. Find the reading on the clock in S opposite the point, in S´, where event 2 occurred.

Notice the symmetry in these drawings. An observer always finds the clocks at rest with respect to him to be synchronized and always finds clocks moving with respect to him to be out of synchronization.

These diagrams help to explain a common misconception about the Lorentz transformation. Suppose we wish to know the time of event 2, as measured in S´. This event occurs at t=0 and the clocks in S are synchronized so they all read t=0 when the event occurs. Furthermore, the clocks in S´ are also synchronized and they all read the same time when the event occurs. Why not pick an arbitrary value of x and carry out the transformation using this x and t=0? This certainly gives the time on the clock in S´ opposite x when the clock in S reads 0.

Here we recall that the transformation equations deal with the coordinates and time of a single event. The event in question here is at x_2, t_2 and not at x, t_2. To be sure, there could be an event at x,t_2 but only in S is it simultaneous with the one at x_2,t_2.

From the viewpoint of an observer in S, the clock opposite x in S´ does not read the same as the clock opposite x_2 in S´ and so does not read the time of the event at x_2. From the point of view of an observer in S´, the clock at x does read 0 but the clock at x_2 does not and so, again, the clock in S´ does not read the time of the event. It is essential that the values of x and t used in the transformation equations be the coordinates and time of the event of interest.

Problem 3.   At time t=0, two events occur. One is at the origin. For each of the following positions of the second event, find the time interval between the events, as measured in the primed frame, which moves in the positive x direction with speed $v=2.4\times10^8$ m/s.

a. $x=1\times10^7$ m.

b. $x=5\times10^6$ m.

c. $x=1\times10^6$ m.

d. $x=5\times10^5$ m.

e. $x=0$.

The relative nature of simultaneity enters into length measurements. Suppose it is desired to measure the length of a rocket ship as it moves by. A standard technique is as follows. Two marks are made on the coordinate axis parallel to the rocket's velocity and at rest with respect to the observer. One mark is at the front of the rocket while the other is at the back. Then the distance between the marks is measured or the difference between their coordinates is calculated. It is clear that the two marks must be made simultaneously so that the measurement does not include the distance moved by the rocket in the interval between the making of the two marks.

Problem 4.   Suppose the S´ frame is attached to a rocket ship traveling at 0.97c in the positive x direction of S. An observer in S simultaneously makes two marks on his x coordinate axis. One, at x_1, is at the back of the rocket and the other, at x_2, is at the front. The length of the rocket is found to be $x_2-x_1=85$ m.

a. Take $x_1=0$, $t_1=0$, $x_2=85$ m, $t_2=0$ and find the coordinates of the two events, as measured in S´. If the marks are placed on both the x and x´ axes, $x_1´$ must be at the back of the rocket and $x_2´$ must be at the front, so $x_2´-x_1´$ is the length of the rocket as measured in S´. This is called the rest length of the rocket and it is the length as measured by a meter stick at rest with respect to the rocket.

b. Start with the Lorentz transformation equations and show that $\Delta x´=\gamma \Delta x$ for two events which are simultaneous in S. Here $\Delta x=x_2-x_1$ and $\Delta x´=x_2´-x_1´$. Calculate γ for $v=0.97c$, then calculate $\Delta x´$ for $\Delta x=85$ m and compare the result with the answer to part a.

c. The moving rocket is found to be shorter than its rest length. This result is quite general and does not depend on which frame is the rest frame of the rocket. Suppose the rocket is at rest in S and an observer in S´ measures its length. Suppose the distance between the marks is 85 m, in S´. Use the inverse Lorentz transformation to find the rest length, now measured in S.

d. Start with the Lorentz transformation equations and show that $\Delta x´=\Delta x/\gamma$ for two events which are simultaneous in S´. Is the result of part c in agreement with this expression?

For parts a and b, the marks are made simultaneously in S but not in S´. From the point of view of an observer in the rocket, frame S moves a distance $v\Delta t´$ during the time interval $\Delta t´$ between the making of the two marks. This observer claims that the mark at the front of the rocket is made first and, before the mark at the back of the rocket is made, the first mark moves with speed v toward the back of the rocket. When the second mark is made, the distance between the marks is clearly less than the length of the rocket.

Problem 5. Consider the measurement process described in problem 4, part a. As the rocket moves in the positive x direction with a speed of 0.97c, an observer in S simultaneously places two marks, one at the front and one at the back of the rocket, on the x axis. Take $x_1=0$, $t_1=0$, $x_2=85$ m, $t_2=0$ to be the coordinates and times, respectively, of the two events.

a. Use the transformation equations to find the time interval $\Delta t´$ between the making of the two marks, as measured in S´. Which mark was made first, according to clocks in S´?

b. Find the x´ coordinate of the mark on the x axis, originally at the front of the
rocket, but find this coordinate at the time the mark at the back of the rocket is
made. To do this, set t´=time mark at back of rocket is made, x=85 m, and solve
for x´ analytically, The program cannot be used.

c. To the value of x´ found in part b, add $|v\Delta t´|$ and compare the result to the rest
length of the rocket.

The next problem gives an example of how the length of a moving object cannot
be measured. Carefully compare the situation given here with that given in problem 4.

Problem 6.   Frame S´ is atached to a rocket traveling at 0.97c in the positive x
direction of frame S. An observer in S´ simultaneously makes two marks. One, at $x_1´$,
is at the back of the rocket and the other, at $x_2´$, is at the front. The rest length
of the rocket is $x_2´-x_1´=300$ m and the marks, when they are made, are placed on both
the x and x´ axes.

a. Take $x_1´=0$, $t_1´=0$. $x_2´=300$ m, $t_2´=0$, and find the coordinates of the two events,
as measured in S. What is the distance between the marks on the x axis, as
measured in S?

b. What is the length of the rocket as measured in S?

If, in one frame, two events occur at the same place but at different times,
they have different coordinates in another frame, moving with respect to the first, and
the time interval between them is longer, as measured in the second frame. The time
interval as measured in the frame for which the events occur at the same coordinate is
called the proper time interval for the events and it is always shorter than the interval measured in any other frame. If the two events occur at the same x´ coordinate,
then the clocks in S´ measure the proper time interval and che time interval as measured
in S is longer. In fact, $\Delta t=\gamma\Delta t´$. This phenomenon is called time dilation. Since the
word dilation means stretching or getting longer, it is an apt name.

The faster the relative speed of the two frames, the greater the dilation.

454

Problem 7. At t=0, a radioactive decay occurs at $x=3\times10^7$ m. 0.5 s later, as measured in S, a second decay occurs at the same place. These events are also viewed from S´, moving with respect to S. For each of the following velocities of S´, find the coordinates and times of the events, as measured is S´.

a. v=0.99c.

b. v=0.95c.

c. v=0.8c.

d. v=0.2c.

Note that, for slow speeds, both the coordinates and times of the events are closer together than for higher speeds but the time interval is longer in S´ than in S, no matter what the speed.

Problem 8. A rocket, traveling at 0.96c, crosses a galaxy in 1.7 yr (5.36×10^7 s), as measured by clocks on board. Frame S is attached to the rocket and frame S´ is attached to the galaxy. The two events considered are the entry of the rocket into the galaxy and its exit from the galaxy.

a. In which frame is the proper time of the trip measured? Multiply or divide by γ, as appropriate, to find the time of the trip in S´.

b. How wide is the galaxy, as measured by observers on the rocket? Is this the rest width? Multiply or divide by γ, as appropriate, to find the width of the galaxy in S´. Divide the length of the trip by the time interval, both as measured in S´. This should yield the speed of the rocket.

c. Check your answers by using the Lorentz transformation equations. Consider two events which occur at opposite ends of the galaxy, separated in time by 5.36×10^7 s in S, and find the coordinates and times in S´. The distance between the events in S is, of course, zero.

A situation quite similar to that described in problem 8 occurs often and can be used to verify time dilation experimentally.

Problem 9. A muon is created in a nuclear reaction, 3000 m above the surface of the earth. It travels at 0.99c toward the surface and, in its rest frame, it lives for 1.5×10^{-6} s before it decays into other particles. Note that, if the time interval were the same in all frames, the muon would travel about 446 m before its decay and so would not reach the earth.

a. Multiply or divide the lifetime by γ, as appropriate, to find the lifetime of the muon, as measured in the rest frame of the earth. How far would the muon travel, if unimpeded, as measured in this frame? Does it reach the earth?

b. The same conclusion must be reached if information given for the rest frame of the muon is used. As measured in the muon's frame, how far away is the earth when the muon is created? In that frame, the earth is moving toward the muon at a speed of 0.99c. How far does the earth move during the lifetime of the muon? Does the earth reach the muon before the particle decays?

c. Check these results by using the Lorentz transformation. The two events are the creation and decay of the muon, a time interval of 1.5×10^{-6} s apart in the muon's rest frame. The distance between them, in the muon's frame, is zero.

It may be that the two events considered do not occur at the same coordinate in either S or S´. Then neither Δt nor $\Delta t´$ is the proper time interval between them and these two quantities are not related to each other by a factor of γ. It may also be that the two events do not occur simultaneously in either S or S´. Then Δx and $\Delta x´$ are not related by the factor γ either.

Problem 10. A space ship travels at 0.96c relative to an earth bound observer. Event 1 occurs at the front of the ship and event 2 occurs 3.2×10^{-6} s later at the back, 200 m away. This time and distance are measured by instruments at rest with respect to the ship. Find the time interval and distance between the two events as measured by instruments at rest with respect to the earth. Show by direct comparison that neither the ratio of the distance intervals nor the ratio of the time intervals equals γ.

For some pairs of events, it is possible to find a frame S´, traveling at less than the speed of light, in which the two events occur at the same time. For other pairs of events, it is possible to find a frame S´, traveling at less than the speed of light, in which the two events have the same x´ coordinate.

Problem 11.   For four pairs of events, the separation in space and in time, as measured in frame S, are given below. For each pair, find the velocity of a frame S′, in which the events occur simultaneously and also find the velocity of a frame S′′, in which the events have the same x coordinate. To do this, solve either Eqn. 23-9 or 23-10 for v. If v is less than c in magnitude, use the program of Fig. 21-1 to verify that $\Delta t'=0$ or $\Delta x''=0$, as appropriate.

a. $\Delta x=5.4\times10^{-7}$ m, $\Delta t=2.5\times10^{-14}$ s.

b. $\Delta x=5.4\times10^{-6}$ m, $\Delta t=1.5\times10^{-14}$ s.

c. $\Delta x=4$ m, $\Delta t=2.5\times10^{-8}$ s.

d. $\Delta x=6$ m, $\Delta t=1.5\times10^{-8}$ s.

The relative nature of simultaneity is sometimes erroneously ignored in order to present a paradox. The paradox is explained when relativistic effects are correctly taken into account. A famous example is presented in the next problem.

Problem 12.   A barn is 30 m long and has doors in both the front and back walls. Both doors are open. A 40 m rod, traveling at v=0.95c, enters the front door and, when its back end is inside the barn, both doors are closed. Since, in the rest frame of the barn, the length of the rod is 40/γ m long, and therefore shorter than 30 m, this can clearly be done with the rod entirely within the barn. Shortly thereafter, of course, the front end of the rod slams through the closed rear door of the barn. For a short time, however, the rod is entirely in the barn with both doors shut. From the point of view of an observer moving with the rod, the rod is 40 m long and the barn is shorter, 30/γ m. It is clearly impossible for the rod to fit in the barn. What sequence of events does this observer see?

a.   Suppose the front door of the barn is at x=0, the rear door is at x=30 m, and the front end of the rod enters the barn at t=0. Calculate the time when the back end of the rod enters the barn and the time when the front end of the rod reaches the back door of the barn. To calculate the first of these times, you must find the length of the rod in the frame of the barn.

b.   Use the Lorentz transformation to find the position and time of the three events in the rest frame of the rod. The events are: the front of the rod enters the barn, the back of the rod enters the barn, and the front of the rod reaches the back of the barn.

c.   Put the events in chronological order according to clocks at rest with respect to the rod.

It is possible to give a nice demonstration of time dilation, in which each observer signals the other as to the time read by his clocks. Each finds that the other's clocks are slow. This is done in the next problem. To solve it, you must make use of the fact that the speed of light is the same in both frames.

Problem 13.   A rocket starts on earth and travels away at a speed of 0.97c. Every hour, an earth bound transmitter sends a radio signal to the rocket. These signals are electromagnetic and travel at the speed of light.

a.   Assume the rocket starts at time t=0 and that the first signal is sent at t=3600 s. Signal n is sent at 3600n (n=1, 2, 3, ...) seconds. Show that signal n is received by the rocket at time t=3600nc/(c-v) seconds and that the distance to the rocket is x=3600ncv/(c-v) when it receives signal n. Find the numerical values of t and x for the first five signals.

b.   Use the Lorentz transformation to find the time t', as measured by the rocket's clock, when each of the first five signals is received.

c.   The observer on the rocket can find the time, according to clocks on the rocket, at which a signal is sent. If t' is the time of reception and t_0' is the time of sending, then the earth-rocket distance at the time of sending is given by $d'=vt_0'=c(t'-t_0')$. Show that $t_0'=t'c/(c+v)$ and use this expression to find the time at which each of the first five signals is sent. According to the rocket clocks, is the earth clock slow or fast?

d.   Now assume the rocket emits a signal every hour, according to the rocket clock. Carry out an analysis similar to the one described in parts a, b, and c. Find the times at which the first 5 signals are received at the earth, then use the inverse Lorentz transformation to find the reception times, according to earth clocks. Finally, use this information to find the sending time for each of the first five signals, as measured by the earth clock. According to the earth clock, is the rocket clock slow or fast?

e.   The results of part c can be obtained by applying the transformation equations to the sending events. These occur at x=0, t=3600n s. The results of part d can be obtained in a similar manner, by applying the inverse transformation to events at $x'=0$, $t'=3600n$ s. Apply the transformation equations to the first signal sent from earth and to the first signal sent from the rocket. Compare the results with those obtained earlier.

23.3 Velocities

Suppose an object moves with velocity u_x in the positive x direction, the velocity being measured by instruments at rest in frame S. If the velocity of the same object is measured by instruments at rest in S´, the result is different, $u_x´$, say. Someone who did not know about special relativity might guess that $u_x´=u_x-v$ where v is the velocity of S´ relative to S. If a man is walking at 5 mph, his velocity relative to a car going 35 mph in the same direction is -30 mph.

Using the Lorentz transformation equations, however, it is easy to see that this method of computing $u_x´$ is in error whenever u_x or v is close to the speed of light. A velocity is the ratio of a distance measurement to a time measurement and the transformation equations can be used to find both in S´.

If a particle has velocity u_x along the x axis, then it goes from x_0 to $x_0+u_x\Delta t$ in the time interval from t_0 to $t_0+\Delta t$. The two events of interest are at the beginning and end of the time interval and they occur at the position of the particle at those times. In frame S´, the first event is at $x_0´,t_0´$ and the second event is at $x_0´+\Delta x´,t_0´+\Delta t´$. The velocity of the particle, as measured in S´, is $\Delta x´/\Delta t´$.

Problem 1.   Suppose a particle, at the origin at time t=0, moves in the positive x direction with velocity $u_x=0.95c$.
a.   Find its position at t=5 s.
b.   Find its velocity in S´, a frame moving at 0.92c in the positive x direction.
To do this, find the coordinate and time, in S´, of each of the two events, then calculate $u_x´=\Delta x´/\Delta t´$. Notice that the velocity in S´ is not 0.03c.

If a particle, moving with velocity u_x in one direction, is viewed by an observer going in the opposite direction, the velocity obtained by the observer is greater than u_x. If the particle velocity and the observer's velocity are both less than the speed of light, then the velocity in the observer's frame is also less than the speed of light.

Problem 2.   In frame S, a particle moves along the x axis with a velocity of 0.96c.
For each of the following values of the velocity of frame S´, find the particle
velocity in that frame.
a.   v=-0.1c.
b.   v=-0.9c.
c.   v=-0.99c.

        If the particle is moving at the speed of light in frame S, then its speed in
frame S´ is also the speed of light.

Problem 3.   In frame S, a particle moves along the x axis with speed c. For each of
the following values of the velocity of frame S´, find the particle velocity in that
frame.
a.   v=-0.1c.
b.   v=-0.9c.
c.   v=+0.9c.

        If the particle moves in the y direction in frame S then, from the viewpoint of
S´, neither the x nor the y component of the velocity vanishes. They are given by
$u_x´=\Delta x´/\Delta t´$ and $u_y´=\Delta y´/\Delta t´$ respectively. According to the Lorentz transformation
equations, $\Delta y´=\Delta y$. However, $u_y´$ is not the same as u_y because $\Delta t´$ is not the same as
Δt.

Problem 4.   A particle moves in the positive y direction of S with velocity u_y=0.92c.
a.   Select an initial position and a time interval. Write down the coordinates of the
      particle at the beginning and end of the interval.
b.   Find the components of the velocity in a frame moving in the positive x direction
      at v=0.9c. Do this by using the transformation equations to find $\Delta x´$ and $\Delta t´$,
      then evaluate $\Delta x´/\Delta t´$ and $\Delta y´/\Delta t´$.
c.   Repeat the calculation of part b for a frame S´, moving in the positive x direc-
      tion with speed 0.95c.

460

d.   Repeat the calculation of part b for a frame S´, moving in the negative x direction with speed 0.9c.

e.   For the S´ frames of parts b, c, and d, find the angle between the particle's direction of motion and the x´ axis.

An algebraic expression for the velocity components in S´ can be obtained from the transformation equations. Event 1 occurs at the beginning of the time interval and event 2 occurs at the end. In S´

$$x_1{´} = \gamma(x_1-vt_1) \qquad \text{and} \qquad x_2{´} = \gamma(x_2-vt_2). \qquad (23\text{-}14)$$

Subtract to find

$$\Delta x´ = \gamma(\Delta x-v\Delta t). \qquad (23\text{-}15)$$

Similarly,

$$\Delta t´ = \gamma(\Delta t-v\Delta x/c^2). \qquad (23\text{-}16)$$

Divide the first of these equations by the second to obtain

$$\frac{\Delta x´}{\Delta t´} = \frac{\Delta x-v\Delta t}{\Delta t-v\Delta x/c^2}. \qquad (23\text{-}17)$$

The left side is $u_x{´}$. Now divide both the numerator and the denominator of the right side by Δt and recognize that $\Delta x/\Delta t=u_x$. The result is

$$u_x{´} = \frac{u_x-v}{1-vu_x/c^2}. \qquad (23\text{-}18)$$

A similar analysis for the y component yields

$$u_y{´} = \frac{\Delta y´}{\Delta t´} = \frac{u_y}{\gamma(1-vu_x/c^2)}. \qquad (23\text{-}19)$$

The z component obeys a transformation equation which is similar to that for the y component:

$$u_z' = \frac{u_z}{\gamma(1-vu_x/c^2)} \quad . \qquad \qquad (23\text{-}20)$$

Problem 5.

a. Write a program which produces u_x' and u_y' when v, u_x, and u_y are entered. Use Eqns. 23-18 and 23-19.

b. Check the equations and the program by using them to rework problems 2 and 4.

Problem 6. In frame S, the velocity of a moving particle is $0.52c\hat{i} + 0.84c\hat{j}$. An observer moves along the x axis and measures the velocity of the particle.

a. Use the program developed in answer to problem 5 to plot u_x' as a function of the observer's velocity. Plot points every $0.1c$ from $v=-c$ to $v=+c$.

b. Use the program of problem 5 to plot u_y' as a function of the observer's velocity. Plot points every $0.1c$ from $v=-c$ to $v=+c$.

In all frames, except those with $v=\pm c$, the speed of the particle is less than the speed of light. At the two extremes, for which the magnitude of u_x' equals the speed of light, u_y' vanishes. The x' component u_x' vanishes for $v=0.52c$. Then the S' frame is moving with a speed equal to u_x.

Problem 7. Two rocket ships are moving relative to an observer at rest in frame S. The first moves in the positive x direction with velocity $u_{1x}=0.94c$, $u_{1y}=0$ and the second moves in the negative x direction with velocity $u_{2x}=-0.9c$, $u_{2y}=0$. They move toward each other. Use the program of problem 5 to answer the following questions.

a. What is the velocity of the second rocket as measured by an observer on board the first?

b. What is the velocity of the first rocket as measured by an observer on board the second?

c. The observer on the first rocket sends a message capsule toward the second rocket. The capsule travels at $0.8c$ relative to the first rocket. At what speed does the second observer see the capsule approach?

d. What is the speed of the capsule relative to S? S has a velocity of $-0.94c$ relative to the first rocket.

462

Problem 8. Two rocket ships are moving relative to an observer at rest in S. The first moves with velocity $u_{1x}=0.94c$, $u_{1y}=0$ and the second moves with velocity $u_{2x}=0$, $u_{2y}=0.9c$. Both rockets start at the origin at t=0.

a. What are the velocity components of the second rocket, as measured by an observer on board the first? What is the speed of the second rocket according to this observer?

b. At t=0.5 s, as measured in S, the observer on the first rocket beams a light signal to the second rocket. What angle does the beam make with the x axis of S? Develop the equation for the path of the beam. In particular, show that

$$x_{10}\sin\alpha = -y_{20}\cos\alpha + (u_{2y}/c)x_{10},$$

where α is the angle between the beam and the x axis, x_{10} is the x coordinate of rocket 1, and y_{20} is the y coordinate of rocket 2, when the beam is sent. Solve for the angle α using a root finding program.

c. At what angle to the x axis of his rest frame must the sender aim the beam? First find the velocity components of the light signal, as measured in S, then transform to the rest frame of the rocket.

Chapter 24

RELATIVISTIC DYNAMICS

.

In this chapter the definitions of relativistic momentum and energy are given and their transformation equations are discussed. Problems involving the integration of Newton's second law are presented and the transformation law for a constant force is discussed. This is meant to illustrate the idea that the laws of physics are the same in all inertial frames. Finally, the conservation laws for momentum and energy are presented in the context of collisions between particles. This material augments the supplementary topics section of PHYSICS and FUNDAMENTALS OF PHYSICS, Chapter III of INTRODUCTION TO SPECIAL RELATIVITY, and Chapter 3 of RELATIVITY AND EARLY QUANTUM THEORY.

24.1   Momentum and Energy

For a high speed particle, the definitions of momentum and energy are different from those used in non-relativistic mechanics. The relativistic definitions are

$$\vec{p} = \frac{m\vec{u}}{(1-u^2/c^2)^{\frac{1}{2}}} \qquad (24-1)$$

and

$$E = \frac{mc^2}{(1-u^2/c^2)^{\frac{1}{2}}} \qquad (24-2)$$

for the momentum and energy respectively. Here m is the mass of the particle, $\vec{u}$ is the velocity of the particle, u is its speed, and c is the speed of light. The velocity components and the speed are measured relative to some frame S. The momentum and energy, given by Eqns. 24-1 and 24-2, are for that frame.

These quantities are defined so that the total momentum and energy of a collection of particles are conserved as they interact, provided no external forces act on the particles. We study collisions in a later section.

The energy, as given by Eqn. 24-2, includes the kinetic energy of the particle but excludes any potential energy. It also includes what is known as the rest energy of

463

464

the particle, a concept which is now discussed.

If the speed of the particle is much less than the speed of light, the denominators of Eqns. 24-1 and 24-2 can be expanded, using the binomial theorem:

$$(1-u^2/c^2)^{-\frac{1}{2}} = (1)^{-\frac{1}{2}} - \frac{1}{2}(1)^{-3/2}(-u^2/c^2) + \frac{3}{8}(1)^{-5/2}(-u^2/c^2)^2 + \ldots$$

$$= 1 + \frac{1}{2}(u^2/c^2) + \frac{3}{8}(u^2/c^2)^2 + \ldots. \qquad (24-3)$$

Since $u/c \ll 1$, each term in the expansion is smaller than the preceding one. The definitions become

$$\vec{p} = m\vec{u} \left(1 + \frac{1}{2}\frac{u^2}{c^2} + \ldots\right) \qquad (24-4)$$

and

$$E = mc^2 + \frac{1}{2}mu^2 + \frac{3}{8}m\frac{u^4}{c^2} + \ldots. \qquad (24-5)$$

The non-relativistic expressions for the momentum, $\vec{p}=m\vec{u}$, and for the kinetic energy, $E=\frac{1}{2}mu^2$, can be recognized in these expressions. For each quantity, the next term has been retained. It is the lowest order relativistic correction and it is useful when one wishes to see if the non-relativistic expression is accurate enough in a given situation. The lowest order correction is calculated and compared with the non-relativistic result. If it is significant, one cannot use the non-relativistic expression.

The expression for the energy contains the term mc^2, which is called the rest energy of the particle. It is the particle energy when $u=0$. Since only energy differences are important, it plays no role if the mass of the particle does not change. There are, however, many situations in which the masses of interacting particles do change and this contribution to the energy is important. Mass is not necessarily conserved in particle interactions.

The energy of the particle, as given by Eqn. 24-2, may be considered to be the sum of two parts, the rest energy and the kinetic energy. The rest energy is mc^2 and the kinetic energy is the difference between the total energy and the rest energy:

$$K = mc^2\left[\frac{1}{(1-u^2/c^2)^{\frac{1}{2}}} - 1\right]. \qquad (24-6)$$

The quantity denoted by m in the definitions of momentum and energy is

sometimes called the rest mass of the particle. This is to distinguish it from another
quantity, called the relativistic mass. We do not make use of this latter concept and
we call m simply the mass of the particle. For a given particle, it has the same value
in all inertial frames of reference.

The energy and momentum are related by

$$E^2 = (pc)^2 + (mc^2)^2 \tag{24-7}$$

where $p^2 = p_x{}^2 + p_y{}^2 + p_z{}^2$ is the square of the magnitude of the momentum. Eqn. 24-7 follows
directly from the definitions, Eqns. 24-1 and 24-2. First take the scalar product of
Eqn. 24-1 with itself to obtain

$$p^2 = \frac{m^2 u^2}{1 - u^2/c^2} , \tag{24-8}$$

then solve for u^2:

$$u^2 = \frac{c^2 p^2}{m^2 c^2 + p^2} . \tag{24-9}$$

The quantity $(1 - u^2/c^2)^{\frac{1}{2}}$ can then be written $mc/(m^2 c^2 + p^2)^{\frac{1}{2}}$ and this is substituted into
Eqn. 24-2 and the result squared to produce Eqn. 24-7.

In another frame S', the momentum and energy are given by similar equations:

$$\vec{p}' = \frac{m\vec{u}'}{(1 - u'^2/c^2)^{\frac{1}{2}}} \tag{24-10}$$

and

$$E' = \frac{mc^2}{(1 - u'^2/c^2)^{\frac{1}{2}}} , \tag{24-11}$$

where $\vec{u}'$ is the particle velocity and u' is its speed, both as measured in S'.

Given the momentum and energy, as measured in one frame, it is possible to
find their values, as measured in another frame. First the velocity components in the
first frame are found. Substitute $(1 - u^2/c^2)^{\frac{1}{2}} = mc/(m^2 c^2 + p^2)^{\frac{1}{2}}$ into Eqn. 24-1 and solve
for $\vec{u}$. The result is

$$\vec{u} = c\vec{p}/(m^2 c^2 + p^2)^{\frac{1}{2}} . \tag{24-12}$$

Then use Eqns. 23-18, 23-19, and 23-20 to find the velocity components in the second frame. Finally, compute the momentum and energy in the second frame, using Eqns. 24-10 and 24-11.

Once the rather tedious algebra is carried out, the results are simply stated. The relationships between the momentum and energy, as measured in S, and those quantities, as measured in S′, are similar to the relationships between coordinates and time in the two frames: p_x replaces x, p_y replaces y, p_z replaces z, and E/c^2 replaces t in the transformation equations. Specifically,

$$\frac{E'}{c} = \gamma(\frac{E}{c} - \beta p_x),$$

$$p_x' = \gamma(-\beta\frac{E}{c} + p_x),$$

$$p_y' = p_y,$$

and
$$p_z' = p_z. \tag{24-13}$$

Here $\gamma = (1-v^2/c^2)^{-\frac{1}{2}}$, and $\beta = v/c$, where v is the velocity of S′ relative to S.

The program of Fig. 21-1 can be used, with the modification given in section 23.2. The Lorentz transformation matrix is the same as was used for the transformation of coordinates and time. The expanded energy-momentum matrix is

$$\begin{pmatrix} E/c & 0 \\ \\ p_x & 0 \end{pmatrix}$$

and, after multiplication by $\underline{\underline{L}}$,

$$\begin{pmatrix} E'/c & 0 \\ \\ p_x' & 0 \end{pmatrix}$$

results.

The following problem is a test of the transformation equations for energy and momentum.

Problem 1. In the laboratory, an electron ($m=9.1096\times10^{-31}$ kg) travels at 0.93c in the x,y plane. Its path makes an angle of 45° with the positive x axis. Its motion is also viewed from the S´ frame, which moves in the positive x direction with a velocity of 0.96c, relative to the laboratory.

a. Find the x and y components of the momentum and the energy of the electron, as measured in the laboratory.

b. Use Eqns. 23-18 and 23-19 to find the components of the electron velocity, as measured in S´.

c. Use Eqns. 24-10 and 24-11 to find the components of momentum and the energy, as measured in S´.

d. Start with the momentum and energy, as measured in S, and use the matrix multiplication program and the Lorentz transformation equations to find the momentum and energy, as measured in S´. Compare the results with those of part d.

Problem 2. A 2.7×10^{-27} kg particle travels along the x axis at a speed of 6×10^{-3}c.

a. Calculate the rest energy, the total energy, and the momentum. For the last two quantities, use the defining equations, Eqns. 24-1 and 24-2.

b. Subtract the rest energy from the total energy and compare the result with $\frac{1}{2}mu^2$.

c. Now suppose the particle moves at 0.92c. Repeat the calculations of parts a and b.

If the S´ frame is attached to the particle, then the momentum in that frame is zero and the energy is the rest energy of the particle. Consider a particle which moves along the x axis of frame S. According to Eqns. 24-13, the momentum in S´ vanishes if

$$v = p_x c^2/E. \qquad\qquad (24-14)$$

This is the velocity of S´ and, since S´ is attached to the particle, it is also the velocity of the particle, as measured in S.

Problem 3. A particle moves in the positive x direction of S. It has momentum 2.7×10^{-18} kg·m/s and energy 8.6×10^{-10} J.

a. Use $E^2=(pc)^2+(mc^2)^2$ to find the mass of the particle.

b.   Use Eqn. 24-14 to find the velocity of a frame S´ which moves with the particle.

c.   Use the matrix multiplication technique to find the momentum and energy of the particle, as measured in S´. Verify that the momentum is zero.

d.   Use $E´=mc^2$ to find the mass of the particle and compare with the answer found in part a.

As the speed of a particle, with mass, increases toward the speed of light, its kinetic energy increases and, for speeds near the speed of light, the increase is dramatic. According to the defining equations, Eqns. 24-1 and 24-2, both the energy and momentum become infinite as u approaches c. The ratio of the energy to the magnitude of the momentum, however, does not blow up, but approaches c in the limit as the speed approaches the speed of light. Eqn. 24-7 predicts that, when pc is much greater than mc^2, the energy is nearly proportional to the magnitude of the momentum and the constant of proportionality is c.

Problem 4.   A 6.5×10^{-29} kg particle travels along the x axis. Use the program of Fig. 2-4 (Appendix C) to make separate plots of the energy, the momentum, and the ratio E/p as functions of the particle speed. Plot points every 0.05c from u=0 to u=c but omit points for which any of the quantities are infinite. On the energy plot, mark the value of the rest energy.

At slow speed, the energy is nearly the rest energy. Added to this is the kinetic energy, which increases as the square of the speed. Close to the speed of light, the energy increases more rapidly with the speed of the particle and becomes infinite at u=c.

The momentum starts at zero and increases linearly with the speed of the particle. At relativistic speeds, it increases more rapidly and becomes infinite at u=c.

The ratio E/p is large near u=0 since p is small. It decreases rapidly since E is nearly constant and p increases. In the relativistic region, E/p approaches c as a limiting value.

These results are important from several standpoints. They form the basis for
the statement that material objects cannot be accelerated from speeds less than the
speed of light to speeds greater than the speed of light. To do so requires that
infinite energy be given to the object. If there are particles which travel faster
than light, they cannot be slowed to speeds less than the speed of light without an
expenditure of infinite energy.

Some particles do travel at the speed of light but have finite energy and
momentum. A particle of light, the photon, is an example. Such particles have zero
mass and, for them, Eqns. 24-1 and 24-2 cannot be used to calculate the momentum and
energy, given the speed. Since both numerator and denominator are zero, the equations
are indeterminate. For these particles, the energy and momentum are still related
by Eqn. 24-7 and E=pc. Eqns. 24-13 for the transformation of the energy and momentum
are also valid.

Problem 5. A photon, when observed with instruments at rest in frame S, has an energy
of 7.2×10^{-19} J. It moves in the x,y plane. Both the x and y components of its momentum
are positive and its path makes an angle of 32° with the x axis of S.
a. What are the cartesian components of its momentum, as measured in S?
b. The energy and momentum of the photon are measured in S´, a frame of reference
which moves with velocity 0.96c in the positive x direction. What are the energy
and momentum in that frame? Use the Lorentz transformation equations.
c. What angle does the photon's path make with the x´ axis?
d. Compute p´c and compare the result with E´.
e. Use the angle calculated in part c to compute the x and y components of the photon
velocity, as measured in S´.

For a system comprised of more than one particle, the total momentum is the
vector sum of the individual momenta and the total energy is the sum of the individual
energies, all measured in the same frame. The total momentum and the total energy, as
measured in another frame, are given by Eqns. 24-13, where $\vec{p}$ and E are now interpreted
as the total momentum and energy, respectively.

For a given system of particles, it is possible to find a frame S´ in which the
total momentum vanishes. This frame is called the center of mass frame, although

perhaps it should more rightfully be called the zero momentum frame or the center of momentum frame.

Eqn. 24-7 is used to define what is meant by the mass of the system. Given the total energy and total momentum, it is solved for m. The result is not necessarily the sum of the masses of the individual particles which make up the system. The mass of the system is the total energy in the center of mass frame, divided by the square of the speed of light.

Problem 6.   Two particles move toward each other along the x axis of frame S. The first particle has a mass of 7.6×10^{-25} kg and moves to the right with a speed of 0.96c while the second particle has a mass of 5.4×10^{-25} kg and moves to the left with a speed of 0.91c.

a.   Find the individual momenta, the individual energies, the total momentum, and the total energy, as measured in S.

b.   Find the velocity of a frame S´ in which the total momentum vanishes.

c.   Use the Lorentz transformation equations to find the momentum and energy of each particle, as measured in the center of mass frame. Sum the individual momenta to find the total momentum and sum the individual energies to find the total energy in S´.

d.   Start with the total momentum and energy, as measured in S and use the Lorentz transformation equations to find their values, as measured in S´. Compare the results with those of part c.

e.   Use Eqn. 24-7 to find the mass of the system. Calculate it twice, once using values of momentum and energy as measured in S and once using values of momentum and energy as measured in S´.

Notice that the mass of the composite system is greater than the sum of the masses of the individual particles which comprise the system. The mass of the system depends on the kinetic energy of the constituent particles.

It is important to distinguish between motion of the individual particles and motion of the system as a whole. When the total momentum vanishes, the system as a whole is said to be at rest even though the individual particles are moving. Any kinetic energy the particles have when the system as a whole is at rest contributes to the internal energy of the system and forms part of the system's rest energy. Since $E=mc^2$ for a system at rest, the mass of the system increases with an increase in the internal energy.

As an example of a system, consider a hydrogen atom, composed of a proton and an electron. The atom is at rest if the total momentum vanishes, but then both particles are moving and have non-vanishing kinetic energy. The system also has a potential energy and both this and the kinetic energy contribute to the rest energy of the atom. For a bound system, such as the hydrogen atom, the energy is less than the total energy the particles would have if they were well separated and at rest. The mass of the atom is less than the sum of the masses of the particles which comprise it.

The difference between the sum of the rest energies of the individual particles and the rest energy of the system is called the binding energy of the system and it represents the minimum energy which must be supplied in order to break the system apart and free the constituent particles.

For the hydrogen atom, the binding energy is extremely small compared to the individual rest energies. The binding energies of atomic nuclei, however, are large and play an important role in many of the phenomena of nuclear physics. Some of these are discussed later.

If the center of mass energy is changed by an amount ΔE, this change is accompanied by a change Δm in the mass and the two changes are related by $\Delta E=(\Delta m)c^2$. It is possible, of course, to increase the energy of the system without changing the internal energy. This occurs when the center of mass is caused to move faster without change in the internal motion relative to the center of mass. This change in energy is not accompanied by a change in mass. The acceleration of a single particle, for example, does not change its rest energy or its mass, as we have used the term.

24.2   Newton's Second Law

Newton's second law, in the form $\vec{F}=m\vec{a}$, does not hold for a particle with speed near the speed of light. The form which is relativistically correct is

$$\vec{F} = d\vec{p}/dt \tag{24-15}$$

where $\vec{p}$ is the relativistic momentum, defined by Eqn. 24-1, and $\vec{F}$ is the force acting on the particle. At low speeds, $\vec{p}=m\vec{u}$ and $\vec{F}=m\vec{a}$.

The laws of physics are the same in every inertial frame. If $\vec{F}=d\vec{p}/dt$ when $\vec{F}$ and $\vec{p}$ are measured in S, then $\vec{F}'=d\vec{p}'/dt$ when $\vec{F}'$ and $\vec{p}'$ are measured in another frame, S´. In S´, the force may not be the same function of the time and coordinates as in S.

Eqn. 24-15 can be solved for the momentum as a function of time in exactly the same ways that the non-relativistic equation $\vec{F}=m\vec{a}$ can be solved for the velocity as a function of time. Some problems can be solved analytically, while others require numerical techniques for solution. The programs given in chapters 6 and 7 can be used and, once $\vec{p}(t)$ has been found, Eqn. 24-12 can be used to find the velocity. Finally, the position of the particle as a function of time can be found by integrating the velocity:

$$x(t) = x_0 + \int_{t_0}^{t} \frac{cp_x}{(m^2c^2+p^2)^{\frac{1}{2}}} \, dt, \tag{24-16}$$

$$y(t) = y_0 + \int_{t_0}^{t} \frac{cp_y}{(m^2c^2+p^2)^{\frac{1}{2}}} \, dt, \tag{24-17}$$

and
$$z(t) = z_0 + \int_{t_0}^{t} \frac{cp_z}{(m^2c^2+p^2)^{\frac{1}{2}}} \, dt. \tag{24-18}$$

Here $\vec{r}_0 = x_0\hat{i}+y_0\hat{j}+z_0\hat{k}$ is the position of the particle at time $t=t_0$.

At high speeds, the difference between the motion that would be produced if $\vec{F}=m\vec{a}$ were true and the actual motion of a particle is, in some cases, quite great. Consider a particle moving along the x axis and acted on by a force in the same direction. If $\vec{F}=m\vec{a}$ were true, the velocity would increase indefinitely, to values well beyond the speed of light. What actually happens is that the momentum continues to increase but the speed approaches the speed of light as a limit. No matter how long the force acts, the speed of the particle never exceeds that limit.

Problem 1.   A 7.6×10^{-25} kg particle starts from rest and is acted on by a constant force of 4.1×10^{-18} N in the positive x direction.

a.   Assume $\vec{F}=m\vec{a}$. Then $u=(F/m)t$. Draw a graph of u as a function of t for the first 150 s of the motion.

b.   Now use $\vec{F}=d\vec{p}/dt$. For this force law, $p=Ft$. Use the program of Fig. 2-4 to plot u as a function of t on the same graph as was used for part a. Use Eqn. 24-12 to evaluate u within the program and plot points for every 5 s from t=0 to t=150 s.

It is clear that, at high speeds, the acceleration is no longer proportional to the force. The greater the speed of the particle, the larger the force needed to produce a given acceleration. Near the speed of light, enormous forces are required to produce small accelerations. The particle is approaching a limit to its speed.

Not only is the acceleration not proportional to the force but, except for special cases, the acceleration is not in the same direction as the force. Differentiate Eqn. 24-1 with respect to time to find

$$\frac{d\vec{p}}{dt} = \frac{m\,d\vec{u}/dt}{(1-u^2/c^2)^{\frac{1}{2}}} + \frac{mu\vec{u}\,du/dt}{c^2(1-u^2/c^2)^{3/2}}\ . \qquad (24\text{-}19)$$

This can be written in terms of the force. Replace $d\vec{p}/dt$ by $\vec{F}$, $u\,du/dt$ by $\vec{u}\cdot d\vec{u}/dt$, and take the scalar product of both sides with $\vec{u}$ to find

$$\vec{F}\cdot\vec{u} = \frac{m\vec{u}\cdot d\vec{u}/dt}{(1-u^2/c^2)^{\frac{1}{2}}} + \frac{mu^2\,\vec{u}\cdot d\vec{u}/dt}{c^2(1-u^2/c^2)^{3/2}}$$

$$= \frac{m\,\vec{u}\cdot d\vec{u}/dt}{(1-u^2/c^2)^{3/2}}. \qquad (24\text{-}20)$$

In the last step, the addition of the two terms was accomplished by writing them both with the common denominator $(1-u^2/c^2)^{3/2}$. Solve Eqn. 24-20 for $\vec{u}\cdot d\vec{u}/dt$:

$$\vec{u} \cdot d\vec{u}/dt = \frac{(1-u^2/c^2)^{3/2}}{m} \vec{F} \cdot \vec{u}. \qquad (24\text{-}21)$$

When this is substituted back into Eqn. 24-19,

$$\vec{F} = \frac{m \, d\vec{u}/dt}{(1-u^2/c^2)^{\frac{1}{2}}} + \frac{\vec{u}}{c^2} \vec{F} \cdot \vec{u} \qquad (24\text{-}22)$$

results.

It is seen that the acceleration is the sum of two vectors, one of which is in the direction of the force and the other of which is along the line of the velocity. If the force is in the x direction, $dp_y/dt = 0$ but du_y/dt does not vanish unless the x or the y component of the velocity vanishes. This result can also be seen directly from Eqn. 24-1. Since u_x changes with time, the speed u, which appears in the denominator of Eqn. 24-1, also changes with time but the y component of the momentum does not. The y component of the velocity, which appears in the numerator must also change with time, and in just the right way to keep the quotient constant.

Problem 2.   Consider the particle and force of problem 1, but now suppose the particle has an initial velocity of 0.9c in the positive y direction. Again $\vec{p} = Ft\hat{i}$. Use the program of Fig. 2-4 to plot u_y as a function of time. Plot points every 5 s from t=0 to t=150 s. In this case, p_y is a constant. Find its value by using Eqn. 24-1. Then evaluate $p^2 = p_x^2 + p_y^2$ for a selected time and substitute into Eqn. 24-12 to find the y component of the velocity.

There are two special cases for which the acceleration is in the same direction as the force. These occur when the force is in the same direction as the velocity and when the force is perpendicular to the velocity. In the first case, the second term of Eqn. 24-22 is in the same direction as the force while, in the second case, that term vanishes.

If a particle is acted on by a force which is constant in magnitude and always directed perpendicular to the velocity, the momentum and velocity remain constant in magnitude and the particle travels in a circular orbit. The acceleration is then u^2/R

where R is the radius of the orbit. The acceleration is directed toward the center of the circular orbit.

This result is used extensively in high energy physics to measure the momenta of particles. The particle is caused to enter a region of uniform magnetic field $\vec{B}$, where the magnitude of the force, given by $q\vec{u}\times\vec{B}$, is perpendicular to the velocity. For simplicity, suppose the velocity and field are also perpendicular to each other. Then, according to Eqn. 24-22,

$$quB = \frac{mu^2/R}{(1-u^2/c^2)^{\frac{1}{2}}}$$

or

$$\frac{mu}{(1-u^2/c^2)^{\frac{1}{2}}} = qRB. \qquad (24\text{-}23)$$

Since the left side of the last equation is just the magnitude of the momentum, this quantity can be found if the charge, magnetic field, and orbit radius are known. The orbit radius is found from measurements of the particle's track in a bubble chamber or on a photographic film.

A force acting on a moving particle may do work and the work results in an increase in the kinetic energy of the particle. The work done by the force $\vec{F}$ is defined by

$$W = \int \vec{F}\cdot d\vec{\ell} \qquad (24\text{-}24)$$

where $d\vec{\ell}$ is an infinitesimal displacement of the particle and the integral sums the contributions of the various segments of the particle's path. Just as in non-relativistic mechanics,

$$W = \Delta K \qquad (24\text{-}25)$$

if W is the total work done by all forces. Here ΔK is the final kinetic energy minus the initial kinetic energy.

To prove this, substitute $d\vec{\ell}=\vec{u}\,dt$ and $\vec{F}=d\vec{p}/dt$ into the definition for the total work. Then

$$W = \int \frac{d\vec{p}}{dt} \cdot \vec{u} \ dt = \int \frac{d\vec{p}}{dt} \cdot \frac{\vec{p}c}{(m^2c^2+p^2)^{\frac{1}{2}}} \ dt = \int \frac{d\vec{p}}{dt} \cdot \frac{\vec{p}c^2}{\left[(pc)^2+(mc^2)^2\right]^{\frac{1}{2}}} \ dt,$$

where the velocity was written as a function of the momentum and both numerator and denominator were multiplied by c. Now $E = \left[(pc)^2+(mc^2)^2\right]^{\frac{1}{2}}$, so

$$\frac{dE}{dt} = \frac{pc^2}{\left[(pc)^2+(mc^2)^2\right]^{\frac{1}{2}}} \ \frac{dp}{dt} = \frac{d\vec{p}}{dt} \cdot \frac{\vec{p}c^2}{\left[(pc)^2+(mc^2)^2\right]^{\frac{1}{2}}} \ .$$

This is just the integrand of the equation for the work. Furthermore the rest energy is constant, so dE/dt=dK/dt and

$$W = \int \frac{dK}{dt} \ dt = K_f - K_i,$$

which is just ΔK.

If the particle moves along the x axis and is subjected to a force in the x direction,

$$W = \int \vec{F} \cdot d\vec{\ell} = \int F \frac{cp}{(m^2c^2+p^2)^{\frac{1}{2}}} \ dt. \qquad (24\text{-}26)$$

Here Eqn. 24-12 was used to substitute for $\vec{u}$ in $d\vec{\ell}=\vec{u} \ dt$. This integral can be evaluated numerically.

Problem 3.   A 5.8×10^{-27} kg particle starts from rest at time t=0 and moves along the x axis, subjected to the force $\vec{F}=3.6\times10^{-18}t^2 \ \hat{i}$ N.

a.   Show that the momentum is given by

$$p(t) = 1.2\times10^{-18}t^3 \ \text{kg}\cdot\text{m/s}.$$

b.   Use Eqn. 24-26 and the program of Fig. 8-1, with N=32, to calculate the work done by the force during the first 10 s of the motion.

c.   Use Eqn. 24-7 to calculate the kinetic energy of the particle at the beginning
     and end of the 10 s interval. Compare the difference with the answer to part b.
d.   Use Eqn. 24-26 and the program of Fig. 8-1, with N=32, to calculate the work done
     by the force during the first 11 s of the motion. Calculate the fractional
     change in the kinetic energy from t=10 s to t=11 s.
e.   Calculate the velocity of the particle at the end of 10 s and at the end of 11 s.
     Calculate the fractional change in the velocity from t=10 s to t=11 s. The value
     for the velocity must be accurate to at least 8 significant figures.

At the end of 10 s, the speed of the particle is close to the speed of light.
During the next second, the acceleration of the particle is small but the fractional
changes in momentum and energy are large.

Consider a force $\vec{F}$ acting on a particle and producing a change in its momentum,
according to $\vec{F}=d\vec{p}/dt$. Both $\vec{F}$ and $d\vec{p}/dt$ are measured by instruments at rest in frame S.
The interaction of the force and particle can be measured by instruments at rest in
another frame S´. In general, the force $\vec{F}´$, as measured in S´, is not the same as $\vec{F}$
nor is the rate of change of momentum $d\vec{p}´/dt´$ the same as $d\vec{p}/dt$. Nevertheless, the law
of motion $\vec{F}´=d\vec{p}´/dt´$ holds in S´.

Since it is possible to find $d\vec{p}´/dt´$ in terms of $d\vec{p}/dt$, we can use the invari-
ance of the law of motion to find the force as measured in S´, in terms of the force
as measured in S.

Suppose that, at time t, the particle is at x,y and has momentum p_x,p_y and that,
at time t+dt, it is at x+dx, y+dy and has momentum p_x+dp_x, p_y+dp_y. Its energy at time
t is E and, at time t+dt, is E+dE. According to the transformation equations

$$dt´ = \gamma(dt - \frac{v}{c^2} dx),$$ (24-27)

$$dp_x´ = \gamma(dp_x - \frac{v}{c^2} dE),$$ (24-28)

and $$dp_y´ = dp_y.$$ (24-29)

Substitute $dx=u_x\,dt$, $dp_x=F_x\,dt$, $dp_y=F_y\,dt$, and $dE=(F_xu_x+F_yu_y)dt$ into these expressions
to find

478

$$dt' = \gamma(1 - \frac{vu_x}{c^2})\ dt, \tag{24-30}$$

$$dp_x' = \gamma(F_x - \frac{vu_x}{c^2} F_x - \frac{vu_y}{c^2} F_y)\ dt, \tag{24-31}$$

and
$$dp_y' = F_y\ dt. \tag{24-32}$$

Now divide dp_x' and dp_y' by dt' to obtain the components of $\vec{F}'$:

$$F_x' = \frac{F_x - (vu_x/c^2)F_x - (vu_y/c^2)F_y}{1 - vu_x/c^2} = F_x - \frac{vu_y/c^2}{1 - vu_x/c^2} F_y \tag{24-33}$$

and
$$F_y' = \frac{F_y}{\gamma(1 - vu_x/c^2)} \tag{24-34}$$

respectively. Similarly, if there is a z component to the force,

$$F_z' = \frac{F_z}{\gamma(1 - vu_x/c^2)} . \tag{24-35}$$

If, in S, the force is in the x direction, then, in S', it is also in the x
direction and has the same value.

Problem 4.   Consider the situation described in problem 1. At t=0, a 7.6×10^{-25} kg
particle starts from rest at the origin and is acted on for 50 s by a constant force of
4.1×10^{-18} N in the positive x direction. These quantities are measured in frame S. The
motion is also viewed from frame S', which moves at 0.95c in the positive x direction,
relative to S.

a.   Find the initial and final momentum and energy of the particle, as measured in S.
For later use, you also need to know the position of the particle when the force
stops acting. Use the program of Fig. 8-1, with N=32, to evaluate

$$x = \int_0^{50} u_x\ dt.$$

The integrand is given by Eqn. 24-12, with Ft substituted for p.

b.   By direct application of the Lorentz transformation equations, find the initial and final values of the momentum and energy, as measured in S´. Also find the time and position of the particle when the force stops acting, again as measured in S´. Use the matrix multiplication program.

c. Integrate the equation of motion for S´ to find the final momentum, as measured in S´: $p´=p_0´+\int F´dt´$. Then use $E´^2=(p´c)^2+(mc^2)^2$ to find the final energy of the particle, as measured in S´. Compare the values of the momentum and energy with those found in part b.

If the force in S is in the y direction, then the force in S´ has non-vanishing x´ and y´ components.

Problem 5. A 7.6×10^{-25} kg particle starts from rest at the origin at time t=0 and is acted on by a constant force of 4.1×10^{-18} N in the positive y direction. The force is given as measured in frame S. The motion is also viewed from the S´ frame, which moves at 0.95c in the positive x direction, relative to S. Plot $F_x´$ and $F_y´$ as functions of time t´. To do this, consider values of the time t, as measured in S, in increments of 5 s. For each value of t, find p_y, then u_y. Use the Lorentz transformation equations to find t´. For the conditions of this problem, x=0, $p_x=0$, and $u_x=0$ for all values of t. Finally, use Eqns. 24-33 and 24-34 to find the force components. Stop when t=100 s.

The force, as measured in S´, is not constant. Both its direction and magnitude change with time. The force in that frame is also velocity dependent. If the velocity of the particle were different, $\vec{F}´$ would be different. This tells us something important about the nature of the force.

The force has a source. There is a group of objects, not described in the problem, which exert a force on the particle and they are arranged so that the force is constant in S. For the sake of argument, we suppose these source objects are at rest in S. For example, the particle might be a charge between two parallel planes, uniformly covered with charge. Such a distribution of charge creates a uniform electric field and exerts a constant force on the particle as long as it remains between the planes.

From the point of view of S´, the source charges are moving in the negative x´ direction with speed 0.95c and the particle experiences a velocity dependent force. For the example of a charge between two planes of charge, the velocity dependent force should be familiar to you.   It is a magnetic force. The planes, with their charge, are moving and this constitutes a current, which gives rise to a magnetic field.

We can take the example one more step. Suppose the situation in S were different. Instead of a particle between stationary planes of charge, suppose the planes are moving in the negative x direction with a speed of 0.95c. The force, as measured in S, for this situation, is exactly the same as the force, as measured in S´, for the original situation and, in particular, there must be a magnetic field and force. We have used relativity to learn something about electromagnetic forces. Electric and magnetic forces are intimately tied together. The magnetic field of a moving charge is related by relativity to its electrostatic field.

The problem, of course, did not necessarily deal with a uniform electric field. Relativity can be used to investigate the mathematical form taken by other kinds of forces. The technique is, in principle, similar to that described in the problem.

24.3   Collisions

In interactions between particles, the total momentum and total energy are both conserved. In other words, the total momentum after the interaction is the same as the total momentum before the interaction and the total energy after the interaction is the same as the total energy before the interaction.

For the conservation laws to be valid, the quantities before and after the inter- action must be measured in the same reference frame, but the laws are valid in all inertial frames. Observers in S and S´ may report different values for the total energy but they both find that quantity to be conserved. A similar statement can be made about the total momentum.

Interactions in one dimension are considered first. Particle 1 moves in the positive x direction and has momentum p_{1b} while particle 2 also moves parallel to the x axis and has momentum p_{2b}, a number which is positive for motion in the positive x direction and negative for motion in the negative x direction. After the interaction, particle 1 has momentum p_{1a} and particle 2 has momentum p_{2a}. The energy of particle 1 is E_{1b} before the interaction and E_{1a} after the interaction while the energy of particle 2 is E_{2b} before the interaction and E_{2a} after the interaction. Generally, the various quantities are known for the time before the interaction and their values after the interaction are sought.

If $p=p_{1b}+p_{2b}$ is the total momentum before the interaction and $E=E_{1b}+E_{2b}$ is the total energy before the interaction, then the conservation laws are

$$p = p_{1a} + p_{2a} \qquad\qquad (24\text{-}36)$$

and

$$E = E_{1a} + E_{2a}. \qquad\qquad (24\text{-}37)$$

The energy of each particle is related to its momentum by $E^2=(pc)^2+(mc^2)^2$ so the energy conservation law can be written

$$E = \left[(p_{1a}c)^2+(m_1c^2)^2\right]^{\frac{1}{2}} + \left[(p_{2a}c)^2+(m_2c^2)^2\right]^{\frac{1}{2}}. \qquad\qquad (24\text{-}38)$$

This and Eqn. 24-36 are two equations which are to be solved for the two unknowns p_{1a} and p_{2a}. Solve Eqn. 24-36 for p_{2a} in terms of p_{1a} and substitute the result into Eqn. 24-38 to find

$$E = \left[(p_{1a}c)^2+(m_1c^2)^2\right]^{\frac{1}{2}} + \left[(p-p_{1a})^2c^2+(m_2c^2)^2\right]^{\frac{1}{2}}. \qquad\qquad (24\text{-}39)$$

This equation can be solved for p_{1a}, then $p_{2a}=p-p_{1a}$ can be used to find p_{2a}.

The binary search program of Fig. 2-3 can be used to find p_{1a}. Take $p_{1a}c$ to be variable and enter two values which straddle the root into X8 and X9 respectively. At line 510, the instruction should read

```
SQRT(X8↑2+(m₁c²)↑2)+SQRT((pc-X8)↑2+(m₂c²)↑2)-(E)→X2
```

The instruction at line 520 is the same, except X9 replaces X8 and X3 replaces X2. The instruction at line 600 is also the same, except X4 replaces X8 and X5 replaces X2. Values of m_1c^2, pc, m_2c^2, and E must be supplied when the machine is programmed. Once the root is found, divide it by c to obtain the value of p_{1a}.

The product $p_{1a}c$ must have magnitude less than E and this can be used as one limit of the search. One solution to Eqn. 24-39 is $p_{1a}=p_{1b}$, an equality which holds if no interaction occurs. This can be taken as the other limit of the search. Be sure, however, that the exact value for p_{1b} is not included in the range of the search.

Problem 1.   Particle 1 has mass 5.2×10^{-27} kg and moves with momentum 4.6×10^{-18} kg·m/s along the x axis. Particle 2 has mass 7.5×10^{-27} kg and moves with momentum -6.3×10^{-18} kg·m/s along the x axis. After the collision, both particles move along the x axis.

a.   Find the total momentum and energy of the two particles.

b.   Use the binary search program to solve Eqn. 24-39 for the momentum of particle 1 after the interaction.

c.   Find the momentum of particle 2 and the energy of each particle after the interaction. As a check, use these values to calculate the total energy after the interaction and compare the result with that of part a.

It is sometimes convenient to solve the problem in the center of mass frame. The velocity of that frame is given by $v = c^2 p/E$ where p is the total momentum and E the total energy in the original frame. In the center of mass frame, the total momentum vanishes and $p_{1a} = -p_{2a}$. The momentum of one of the particles is then the solution to

$$\left[(p_{1a}c)^2 + (m_1 c^2)^2\right]^{\frac{1}{2}} + \left[(p_{1a}c)^2 + (m_2 c^2)^2\right]^{\frac{1}{2}} - E_c = 0 \qquad (24\text{-}40)$$

where E_c is the total energy in the center of mass frame. In this equation, the values of the momenta used are those appropriate for the center of mass frame. One of the solutions is $p_{1a} = p_{1b}$ and the other is $p_{1a} = -p_{1b}$ where p_{1b} is the momentum of particle 1 in the center of mass frame, before the interaction. The first solution is not of interest since it gives the original momentum, before the interaction. So $p_{1a} = -p_{1b}$ and $p_{2a} = p_{1b}$. The final momenta, in the original frame, can be found by carrying out the inverse Lorentz transformation.

Problem 2.   The interaction is the same as in problem 1. Particle 1 has mass 5.2×10^{-27} kg and moves with momentum 4.6×10^{-18} kg·m/s along the x axis. Particle 2 has mass 7.5×10^{-27} kg and moves with momentum -6.3×10^{-18} kg·m/s along the x axis. After the collision, both particles move along the x axis.

a.   Use the results of part a, problem 1 to find the velocity of the center of mass frame. Then use matrix techniques to find the momentum and energy of each particle in that frame.

b. In the center of mass frame let $p_{1a}=-p_{1b}$, $p_{2a}=p_{1b}$, carry out the inverse transformation, and compare the answers with those of problem 1. You should recognize
that $E_{1a}=E_{1b}$ and $E_{2a}=E_{2b}$ in the center of mass frame.

The collision described above is an elastic collision. Not only is the total
energy conserved but, because total mass and rest energy are also conserved, so is the
total kinetic energy. Kinetic energy is transferred from one particle to the other,
but the total is the same after the collision as it was before.

This is not typical of most interactions between fundamental particles. In
most cases, energy changes form, from rest energy to kinetic or vice versa. This means
that the masses of the particles leaving the interaction are different from the masses
of the particles which enter. The identities of the particles are changed by the interaction. The old particles disappear and new particles take their places. In addition,
the number of particles which leave the interaction may be different from the number
which enter. Such interactions are termed inelastic.

An example of an inelastic interaction occurs when two particles form a single
particle.

Problem 3.   Particle 1 has mass 5.2×10^{-27} kg and moves with momentum 6.1×10^{-18} kg$\cdot$ m/s
along the x axis. Particle 2 has mass 7.5×10^{-27} kg and moves with momentum
-3.1×10^{-18} kg$\cdot$m/s along the x axis.
a. Find the energy of each particle, the total momentum, and the total energy.
b. During the interaction, a single particle is formed. Find its mass.
c. Find the total rest energy and total kinetic energy after the interaction and
    compare them to the total rest energy and total kinetic energy, respectively,
    before the interaction. Is mass created or destroyed during the interaction?

Conservation of energy and momentum are used to find the masses of particles
produced in interactions. Suppose particles 1 and 2 interact. They disappear and, in
their place, particles 3 and 4 are produced. All particles move along the x axis.
Then $p=p_{1b}+p_{2b}$, $E=E_{1b}+E_{2b}$, Eqn. 24-39 is replaced by

$$E = \left[(p_{3a}c)^2 + (m_3 c^2)^2\right]^{\frac{1}{2}} + \left[(p - p_{3a})^2 c^2 + (m_4 c^2)^2\right]^{\frac{1}{2}} , \qquad (24-41)$$

and $p_{4a} = p - p_{3a}$ is used to calculate the momentum of particle 4 after the interaction.

The momenta of the particles after the interaction are found experimentally, if they are charged, by measuring the radii of their orbits in a known magnetic field. Eqn. 24-41 is then used to calculate the mass of one of the particles.

Problem 4. Particle 1 has mass 6.3×10^{-28} kg and moves along the x axis with momentum 3.2×10^{-18} kg·m/s. Particle 2 has mass 3.7×10^{-28} kg and moves along the x axis with momentum -5.9×10^{-18} kg·m/s. Two particles emerge from the interaction. One has mass 4.5×10^{-28} kg and momentum -1.3×10^{-18} kg·m/s. Use the binary search program to solve Eqn. 24-41 for the mass of the other particle, then find its momentum.

Particle decay is the inverse process to the coalescence of two particles to form a single particle. A single particle may decay into two or more particles with energy and momentum again conserved in the process. No matter what the kinetic energy of the original particle, the sum of the masses of the resultant particles can be no greater than the original mass. Consider the decay in the rest frame of the original particle. Since energy is conserved, the sum of the rest energies after the decay can be no greater than the original rest energy. If the resultant particles have non-vanishing kinetic energy, the sum of the rest energies is less. Since the rest energies are the same in every frame, the inequality is true even for frames in which the original particle is not at rest.

Problem 5. A 1.98×10^{-27} kg particle, at rest, decays into two particles, one with 1.67×10^{-27} kg mass and the other with 2.48×10^{-28} kg mass. Find the momenta and energies of the decay products. Solve Eqn. 24-41 numerically, using the binary search program. Assume the heavier product particle moves in the positive x direction.

Most interactions take place in two dimensions. Fig. 24-1 shows particles 1 and 2 entering the interaction along the x axis and particles 3 and 4 leaving the interaction in other directions.

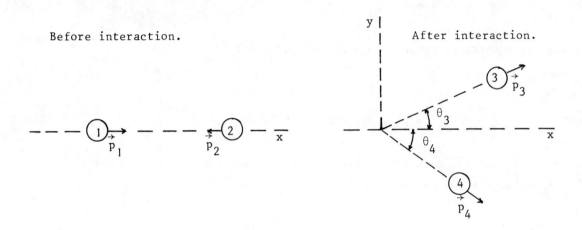

Figure 24-1.   The interaction between two particles. Two different particles are
produced and they move away in the x,y plane.

Each component of the momentum is conserved and the conservation law produces
the two equations

$$p = p_3\cos\theta_3 + p_4\cos\theta_4 \qquad\qquad (24\text{-}42)$$

and
$$0 = p_3\sin\theta_3 - p_4\sin\theta_4, \qquad\qquad (24\text{-}43)$$

where p is the x component of the total momentum, p_3 is the magnitude of the momentum of
particle 3, and p_4 is the magnitude of the momentum of particle 4. The y component of
the total momentum vanishes.

We suppose the total momentum p and the total energy E are given or can be
calculated from given quantities and we solve for the momenta after the interaction as
functions of θ_3. This corresponds to the usual experimental procedure. One of the
particles (3, say) is detected and the angle between its momentum and the direction of
the total momentum is measured.

The angle θ_4 is eliminated from Eqns. 24-42 and 24-43. Write

$$p_4 \cos\theta_4 = p - p_3 \cos\theta_3 \qquad\qquad (24\text{-}44)$$

and
$$p_4 \sin\theta_4 = p_3 \sin\theta_3, \qquad\qquad (24\text{-}45)$$

then sum the squares of these equations to find

$$p_4{}^2 = (p - p_3 \cos\theta_3)^2 + p_3{}^2 \sin^2\theta_3$$
$$= p^2 - 2pp_3 \cos\theta_3 + p_3{}^2. \qquad\qquad (24\text{-}46)$$

Here the square of the quantity in the parentheses was explicitly written out and $\cos^2\theta_3 + \sin^2\theta_3 = 1$ was used.

The total energy is the sum of the individual energies after the interaction, so

$$E = \left[(p_3 c)^2 + (m_3 c^2)^2\right]^{\frac{1}{2}} + \left[(p_4 c)^2 + (m_4 c^2)^2\right]^{\frac{1}{2}}$$
$$= \left[(p_3 c)^2 + (m_3 c^2)^2\right]^{\frac{1}{2}} + \left[(p^2 - 2pp_3 \cos\theta_3 + p_3{}^2)c^2 + (m_4 c^2)^2\right]^{\frac{1}{2}}. \qquad (24\text{-}47)$$

This equation is to be solved for p_3, then Eqn. 24-46 is solved for p_4. Finally, Eqns. 24-44 and 24-45 can be solved for θ_2, if required.

The combination $p_3 c$ is used as the variable in the binary search program and values which straddle the root are entered into X8 and X9. The instructions at line 510 should be

$$\text{(pc)} \uparrow 2 + \text{X8} * (\text{X8} - 2 * (\text{pc}) * \text{COS}(\theta_3)) \rightarrow \text{X10}$$

$$\text{SQRT}(\text{X8} \uparrow 2 + (m_3 c^2) \uparrow 2) + \text{SQRT}(\text{X10} + (m_4 c^2) \uparrow 2) - (\text{E}) \rightarrow \text{X2}$$

First $(p_4 c)^2$ is calculated and stored in X10, then E is subtracted from the expression for the total energy after the interaction. The instructions at line 520 are the same except X9 replaces X8 and X3 replaces X2. The instructions at line 600 are again the same except X4 replaces X8 and X5 replaces X2.

Problem 6. Particle 1 has mass 2.5×10^{-27} kg and moves along the x axis with momentum 7.3×10^{-18} kg·m/s. Particle 2 has mass 4.6×10^{-27} kg and is at rest. Two particles emerge from the interaction. Particle 3 has mass 3.6×10^{-27} kg and particle 4 has mass 4.1×10^{-27} kg. For each of the following values of θ_3, find the momentum and energy of each of the product particles.

a. $\theta_3 = 20°$.
b. $\theta_3 = 40°$.
c. $\theta_3 = 60°$.

Chapter 25
THE PHOTON

The conservation laws for energy and momentum are applied to three situations involving the interaction of light and matter: the photoelectric effect, the Compton effect, and electron-positron annihilation. In each case, the results depend on the assumption that light is composed of particles. As an introduction, the relationship between the particle and wave descriptions of light is discussed and the Doppler shift is examined in terms of particle properties. Use of the programmable calculator allows the relativistic conservation laws to be solved in a straightforward manner, even when the physical situation is more complicated than is usually considered in introductory texts. These topics are meant to be studied in association with Chapter 49 of PHYSICS, Chapter 42 of FUNDAMENTALS OF PHYSICS, Chapter 5 of RELATIVITY AND EARLY QUANTUM THEORY, or similar material in other texts.

25.1   Light Quanta

For electromagnetic radiation of moderate to high intensity, the energy carried by the radiation is nearly proportional to the square of the wave amplitude. This relationship was used in previous sections to find the intensity pattern when light is diffracted by slits. When the intensity is reduced sufficiently, however, this relationship no longer holds and the discrepancy causes us to make a profound change in our concept of the nature of electromagnetic radiation.

There is a great deal of experimental evidence that energy is carried by radiation in bits or particles, called photons. For monochromatic light, each particle carries a specific amount of energy and when the radiation interacts with matter, part or all of that energy is transferred to the matter.

There are important differences between the transfer of energy via a wave and via particles. For a wave, the energy is distributed throughout the region of space where the wave exists while, for a particle, the energy is at the location of the particle. Consider, for example, electromagnetic radiation incident on an electron. If energy were transported by a wave, the electron could absorb, at most, only that energy in the part of the wave which strikes it. Because of the small area presented by the electron, it would take considerable time for it to receive appreciable energy.

On the other hand, the particle model of light suggests that the energy is not distributed continuously over the region of the wave. When the photon-electron interaction occurs, the electron can receive nearly all the energy of the photon and it can receive it instantaneously. The electron does not receive energy continuously but in bursts, as it interacts with incoming photons.

In the wave theory of radiation, the intensity is taken to be the amount of energy which crosses a unit area per unit time and this is proportional to the square of the wave amplitude. In the particle theory, the intensity for monochromatic radiation is given by the number of photons which cross a unit area per unit time multiplied by the energy of one of any one of the photons.

Photons also account for the transfer of momentum. In the particle theory, the momentum obtained per unit time by an electron in an electric field is not given by the product of the charge and electric field. Rather, it is the momentum transferred by an electron-photon interaction multiplied by the number of photons which interact with the electron per unit time. The force is not continuous but occurs discontinuously, when a photon strikes the electron.

Even though energy and momentum are transmitted via particles which move from place to place, there is a wave associated with the radiation. In the modern theory, however, the wave is not interpreted as a propagating force field.

In order to understand the role played by the wave in the theory, it is necessary to know one other characteristic of photons. The motion of any selected photon is not determined by the initial conditions and forces, as it would be if the photon obeyed the laws of classical mechanics. If an experiment consists of the emission of a photon and the subsequent detection of its position and if the same experiment is repeated many times with the same initial conditions, there will be many different outcomes.

What is predictable, given the initial conditions, is the probability that a selected outcome occurs. In the limit of a large number of experiments, this probability is the ratio of the number of times the selected outcome occurs to the total number of trials.

The probability that the photon is in a small region of space in the neighborhood of a given point is proportional to the square of the amplitude of the wave in that region. This is the connection between the photon and the wave. When we calculated the wave intensity on a screen opposite an opaque barrier with slits, we were actually calculating a quantity proportional to the probability that a photon strikes the screen in the region considered.

If the light intensity is high so that many photons strike the screen, the probability is very nearly proportional to the light intensity. At extremely low intensities, however, only a few photons strike the screen and the intensity is quite different from the probability. A given photon strikes somewhere on the screen and, at that place, there is light. Other places are dark. The intensity pattern is not continuous but consists of bright spots where photons strike.

A given photon may strike almost anywhere. It is more likely that it strikes where the probability, as determined by the wave amplitude, is high than where the probability is low, however.

As an example, consider plane waves of light incident on a single slit in an otherwise opaque barrier. See Fig. 22-3. If the viewing screen is far from the slit, the square of the wave amplitude, averaged over a cycle, is proportional to $\sin^2\beta/\beta^2$ where $\beta=(ka/2)\sin\theta$. Here θ is the angle between the ray from the center of the slit to the observation point and the x axis, a is the slit width, and k is the propagation constant for the wave. This quantity is not to be interpreted as the intensity of light. It is, instead, proportional to the probability that a photon strikes the screen in the neighborhood of the ray. More precisely, if dy is an infinitesimal width of screen, then

$$dP = A(\sin^2\beta/\beta^2)\ dy \qquad\qquad (25\text{-}1)$$

is the probability the photon hits the screen in the region from y to y+dy, at the angle θ above the x axis. Here A is a number which is chosen so that the probabilities for all portions of the screen sum to 1.

It is impossible to predict where any particular photon will hit the screen but, if a great many photons are observed, the fraction which hit the screen in a particular region of width dy is nearly equal to the probability given by Eqn. 25-1.

The following is an example to demonstrate the relationship between the wave and the particle motion. It is a calculator simulated experiment in which photons go through a single slit and hit a screen. The calculator is used to determine where each photon hits the screen and you can graphically watch the intensity pattern build up as more photons reach the screen.

To work the problem, you will need a set of integers chosen at random from the set between 0 and 1000, inclusive. Some machines have the capability to produce such a set of numbers. If yours does not, use the set given in Appendix A. When a ramdom number is called for, start anywhere in the list, but from then on, take them in order.

Problem 1.   Light with wavelength 5×10^{-7} m illuminates a 1×10^{-4} m wide slit in a barrier and the intensity pattern is viewed on a screen 1 m behind the slit. The pattern is obtained by means of a series of photon detectors. Each is 1×10^{-3} m wide and registers whenever a photon enters it. These detectors are placed with their centers at $y=0.5\times10^{-3}$ m, 1.5×10^{-3} m, 2.5×10^{-3} m, ..., and 9.5×10^{-3} m, respectively. Here y is the distance from the x axis. See Fig. 22-3. Use the calculator to find where each photon hits the screen. Select an integer randomly from the set 0 to 1000, inclusive, then use the following table to determine which detector receives the photon.

Random Number	Detector Center
0 – 403	0.5×10^{-3} m
404 – 711	1.5×10^{-3} m
712 – 883	2.5×10^{-3} m
884 – 943	3.5×10^{-3} m
944 – 950	4.5×10^{-3} m
951 – 954	5.5×10^{-3} m
955 – 970	6.5×10^{-3} m
971 – 988	7.5×10^{-3} m
989 – 998	8.5×10^{-3} m
999 – 1000	9.5×10^{-3} m

Draw a histogram. Along the horizontal axis of the graph, mark the detector centers. Then, as you consider each photon, place a dot above the detector which receives it.

492

Use uniform spacing between dots so that, for each detector, the dots form a vertical line with length proportional to the number of photons which have entered that detector. Continue until you have considered 250 photons.

This exercise simulates very closely what actually occurs experimentally. There is no way of predicting where any given photon will strike the screen. The detectors are activated in an entirely random manner but some have a higher probability of receiving a photon than others and this probability is determined by the square of the wave amplitude.

If only the first few photons are considered, the intensity pattern on the screen consists of just a few spots scattered here and there. But as more photons are considered, the pattern looks more like the single slit intensity pattern seen in the laboratory or plotted in problem 1 of section 22.2. To obtain closer agreement, consider a larger number of photons.

The distribution of numbers given in the table is designed to reproduce the probability distribution produced by the waves. The probability distribution is such that, in the limit of a large number of photons, 0.403 of them hit between y=0 and $y=1\times10^{-3}$ m, 0.308 of them hit between $y=1\times10^{-3}$ m and $y=2\times10^{-3}$ m, and so on for the other detectors. These numbers are found by evaluating the integrals

$$\int_{0}^{1\times10^{-3}} (\sin\beta/\beta)^2 \, dy, \qquad \int_{1\times10^{-3}}^{2\times10^{-3}} (\sin\beta/\beta)^2 \, dy, \qquad (25\text{-}2)$$

and similar integrals for the other detectors. Finally, the constant A is chosen so that

$$A \int_{0}^{1\times10^{-2}} (\sin\beta/\beta)^2 \, dy = 1. \qquad (25\text{-}3)$$

In the above problem, we considered only those photons which hit the viewing screen between y=0 and $y=10\times10^{-3}$ m. The pattern over a wider range, including negative values of y, can be obtained by adding more detectors. The value of A must then be recalculated.

Interpretation of interference effects must be made in terms of the probability of finding the photon in a selected region of space. When two or more waves are present, the distribution of probability is different than it would be if only one of the waves were present. The probability is not the sum of the probabilities found by considering the individual waves. To find the total probability when several waves are present, the wave functions are first added, then the sum is squared. This procedure allows for both constructive and destructive interference, phenomena which do occur.

In general, the waves are called probability density amplitude waves. The word amplitude occurs in the title because the wave functions must be squared to find the probability. The word density is used since the square of the amplitude gives the probability per unit volume or, in the case of the single slit problem above, the probability per unit length.

Certain characteristics of the wave and the particle are linked. In particular, the frequency of the wave is related to the energy of the particle and the wavelength of the wave is related to the momentum of the particle.

For monochromatic radiation, the energy E of a single photon is related to the frequency ν of the wave by

$$E = h\nu \qquad\qquad (25\text{-}4)$$

where $h = 6.6262 \times 10^{-34}$ J·s. This universal constant, known as Planck's constant, is central to modern quantum theory and it is found in most of the equations of quantum mechanics.

The photon also carries momentum and, for monochromatic plane waves, the momentum is in the direction of propagation for the wave. It is related to the propagation constant of the wave by

$$\vec{p} = h\vec{k}/2\pi \qquad\qquad (25\text{-}5)$$

where $\vec{k}$ has the magnitude $2\pi/\lambda$ and is in the direction of propagation. The magnitude of the momentum is given by

$$p = h/\lambda. \qquad\qquad (25\text{-}6)$$

494

Since $\lambda\nu=c$, it follows from Eqns. 25-4 and 25-6 that $E=pc$. Since, in general, $E^2=(mc^2)^2+(pc)^2$, this relationship is indicative of a particle with zero mass.

Given E and p, as measured in some reference frame, the values E' and $\vec{p}'$, as measured in another frame, can be found by means of the Lorentz transformation equations, Eqns. 24-13.

Although the speed of light is the same in all frames of reference, the wavelength and frequency of a light wave are different in different frames. This dependence of the frequency on the relative motion of the source and observer is known as the Doppler effect and it can be demonstrated by means of the Lorentz transformation equations. This effect, however, does not depend on the particle nature of light. Although the quantitative details are different, it occurs for any wave and, for example, can be heard in the increased frequency of a train whistle as the train approaches. It is presented here as an example of one use of the equations for the energy and momentum of a photon in terms of the frequency and wavelength of the wave.

Problem 2.   A source, when at rest, emits light with wavelength 5×10^{-7} m.
a.   Find the frequency of the light wave, the energy of a photon of this light, and the magnitude of the momentum of a photon of this light.
b.   The light is viewed by an observer moving at 0.97c toward the source. Use matrix multiplication techniques to find the photon energy and momentum in the observer's frame, then use $E'=h\nu'$ and $p'=h/\lambda'$ to find the frequency and wavelength of the light wave in the observer's frame. Verify that $E'=p'c$ and $\lambda'\nu'=c$.
c.   Repeat the calculation of part b for an observer moving away from the source at the speed 0.97c.
d.   Repeat the calculation of part b for an observer moving away from the source at the speed 1×10^{-3}c.
e.   Suppose that light with wavelength 5×10^{-7} m travels in the positive y direction. It is viewed by an observer moving at 0.97c in the positive x direction. Find the energy and the components of the momentum of a photon, in the observer's frame. Use the Lorentz transformation equations, then $E'=h\nu'$ and $p'=h/\lambda'$ to find the frequency and wavelength in the observer's rest frame. Verify that $E'=p'c$ and that $\nu'\lambda'=c$.

When the observer moves toward the source, as in part b of the above problem, the frequency is higher and the wavelength is less than when he is stationary. When the observer moves away from the source, as in part c of the above problem, the frequency is less and the wavelength greater. There is also a transverse Doppler shift: a frequency change occurs for an observer who moves in a direction perpendicular to the direction of wave propagation.

The Doppler effect is used by astronomers to measure the component of a star's velocity away from the earth. The frequency of the light in the frame of the star is known if the light can be identified as characteristic radiation from a known atom, such as hydrogen. The results of such measurements lead astronomers to believe the universe is expanding.

We now go on to examine some situations connected with the experiments which helped establish the photon theory of light. The key to all the problems which follow is that the photon is treated just like any other particle. It undergoes collisions and its energy and momentum change. This means that the frequency and wavelength of the wave change. The underlying phenomenon is the interaction between a photon and other particles.

25.2   The Photoelectric Effect

One phenomenon which supports the photon model of light is the photoelectric effect. A sample of material is illuminated by light, the light is absorbed by particles of the material, and the energy and momentum of the incident light are transferred to those particles. As we shall see, most of the original energy is received by electrons.

In many cases of interest, electrons are ejected from the material following the interaction. For the typical experimental arrangement, the region outside the material is evacuated and the kinetic energy of the most energetic electron is measured. Experimental results are explained by considering the interaction to be one in which single photons are absorbed by single electrons.

If the intensity of the light is increased without changing the frequency, the number of electrons emitted increases but the kinetic energy of the most energetic electron remains the same. The number of photons in the radiation is increased, more interactions take place, and more electrons are ejected. However, the energy transferred in any single photon-electron interaction does not change if the frequency of the light is unaltered.

If, on the other hand, the frequency of the light is increased, it is found that the kinetic energy of the most energetic electron increases. The electron receives more energy from the photon. According to the photon model, $E=h\nu$, so each photon has more energy when the frequency of the light is increased.

When the photoelectric effect occurs, the photon is completely absorbed and its energy and momentum go to the electron and other particles in the material. The presence of another particle is important. It is impossible for a free electron to accept all the energy and momentum of the photon. This is easy to understand. In the reference frame in which the electron is initially at rest, its momentum is $p=h\nu/c$ after the absorption of a photon. If energy is to be conserved, it must be that $mc^2+h\nu=\left[(mc^2)^2+(pc)^2\right]^{\frac{1}{2}}$. This equation is clearly inconsistent with momentum conservation. The electron must be able to exchange energy or momentum with another particle during the photon absorption process.

Generally, the other "particle" involved is much more massive than the electron. It might be an atomic nucleus, a molecule, or a large collection of atoms, bound together in a solid. In such cases, the electron receives nearly all the energy of the photon and the primary role of the other particle is to receive momentum.

In order to study the photoelectric effect, we consider the following situation: the electron, with mass $m=9.1095\times10^{-31}$ kg, and another particle, with mass M, are initially at rest. A photon of light with frequency ν is incident in the positive x direction. After the interaction, the electron has momentum $\vec{p}$ and travels along a line which is at the angle θ_1 above the x axis. The other particle has momentum $\vec{P}$ and travels along the line which is at the angle θ_2 below the x axis.

This situation is somewhat unrealistic. To be sure, it is always possible to find a reference frame in which the electron is initially at rest. In this frame, however, the more massive particle is not at rest and its initial momentum should be taken into account. It is generally small, however, and we ignore it for purposes of this introduction to the effect.

The geometry is somewhat similar to that of Fig. 24-1. Particle 1 is the photon, particle 2 is the electron, particle 3 is also the electron, and particle 4 is the third particle. The third particle exists before the interaction but it is not shown then. The analysis is similar to that which follows the diagram. Conservation of momentum and energy require that

$$h\nu/c = p \cos\theta_1 + P \cos\theta_2, \qquad (25-7)$$

$$0 = p \sin\theta_1 - P \sin\theta_2, \qquad (25-8)$$

and

$$Mc^2 + mc^2 + h\nu = E_e + E_M + \phi. \qquad (25-9)$$

Here $h\nu$ has been used for the photon energy and $h/\lambda = h\nu/c$ has been used for the photon momentum. E_e and E_M are the energies of the electron and the more massive particle, respectively, after ejection of the electron from the material. The energy equation also contains the term ϕ, which represents the energy required to free the electron from the other particle and, if it is in a solid, to move it through the surface. For the electron with the greatest kinetic energy after emission, this term is called the work function. It is positive.

The energy and momentum of the electron are related by $E_e = \left[(mc^2)^2 + (pc)^2\right]^{\frac{1}{2}}$. In many situations of practical interest, this relationship is hard to use since pc is much less than mc^2 and significance is lost in the evaluation of Eqn. 25-9. E_e is very nearly the same as mc^2 and when it is cancelled against the rest energy on the left side of Eqn. 25-9, it is only a small residue which is important.

The electron momentum, in many cases, is small enough that the full relativistic treatment is not required. The binomial theorem is used to expand the expression for the energy and the first three terms are retained:

$$E = \left[(mc^2)^2 + (pc)^2\right]^{\frac{1}{2}} = (mc^2) + \frac{(pc)^2}{2mc^2} - \frac{(pc)^4}{8(mc^2)^3} + \ldots \qquad (25-10)$$

The first term is the rest energy, the second term is the non-relativistic kinetic energy $(\frac{1}{2}p^2/m = \frac{1}{2}mv^2)$, and the third term gives the lowest order relativistic correction to the energy. It is this expression and a similar expression for the energy of the other particle which are used in Eqn. 25-9. This equation then becomes

$$\frac{(pc)^2}{2mc^2}\left[1 - \left(\frac{pc}{2mc^2}\right)^2\right] + \frac{(Pc)^2}{2Mc^2}\left[1 - \left(\frac{Pc}{2Mc^2}\right)^2\right] + \phi - h\nu = 0. \qquad (25-11)$$

For the problems to be considered, $h\nu$, ϕ, and θ_1 are given; p, P, and θ_2 are unknown. A numerical procedure which can be used to solve for the unknowns is now developed.

Eqn. 25-7 is solved for $\cos\theta_2$, Eqn. 25-8 is solved for $\sin\theta_2$, then the resulting expressions are squared and added to eliminate θ_2. The trigonometric identity $\sin^2\theta_2 + \cos^2\theta_2 = 1$ is used. The result is solved for $(Pc)^2$:

$$(Pc)^2 = (h\nu)^2 - 2(pc)(h\nu)\cos\theta_1 + (pc)^2. \qquad (25\text{-}12)$$

This expression is substituted into Eqn. 25-11 and the result is solved for p. Once p is found, Eqn. 25-12 is solved for P. An expression for θ_2 can be found by dividing $\sin\theta_2$, from Eqn. 25-8, by $\cos\theta_2$, from Eqn. 25-7. It is

$$\theta_2 = \tan^{-1} \frac{pc\,\sin\theta_1}{h\nu - pc\,\cos\theta_1}. \qquad (25\text{-}13)$$

The binary search program of Fig. 2-3 can be used to find the root of the function on the left side of Eqn. 25-11. It is worthwhile to use the product pc as the variable and two values which straddle the root are placed in X8 and X9 respectively. The function is evaluated at line 510 by means of the instructions

```
(hν)↑2+X8*(X8-2*(hν)*COS(θ₁))→X10
```

```
(X8↑2/(2*(mc²)))*(1-(X8/(2*(mc²)))↑2)
        +(X10/(2*(Mc²)))*(1-(X10/(2*(Mc²))↑2))+(φ)-(hν)→X2
```

In the first instruction, $(Pc)^2$ is calculated and placed in X10, while, in the second, the left side of Eqn. 25-11 is evaluated.

Similar instructions are programmed at lines 520 and 600. At these lines, either X9 or X4 replaces X8 and the result is placed in either X3 or X5, as appropriate. The correct replacements to be made should be obvious from the flow chart. The root, when found, must be divided by c to obtain the electron momentum p. The quantities $h\nu$, θ_1, mc^2, Mc^2, and ϕ must be supplied at the time of programming. If your machine has sufficient memory, it is worthwhile to store these quantities, then recall them as needed.

The first problem deals with a photoelectric effect experiment in which light is incident on a single atom and one of its electrons is ejected.

Problem 1. A potassium atom has a mass of 6.47031×10^{-26} kg and its outer electron has a binding energy of about 6.92×10^{-19} J. The outer electron absorbs a photon and is ejected in the direction opposite to that of the incident photon. Make the approximation that both the atom and its outer electron are initially at rest.

a. For each of the following frequencies of light, find the momentum and kinetic energy of both the ejected electron and the recoiling potassium atom. Obtain 5 significant figures for the root pc.

   i. $\nu = 2 \times 10^{15}$ Hz.
   ii. $\nu = 4 \times 10^{15}$ Hz.
   iii. $\nu = 6 \times 10^{15}$ Hz.

b. For each frequency, calculate $h\nu - \phi$ and compare the value with the kinetic energy of the ejected electron. Verify that the kinetic energy is proportional to ν.

c. By finding the difference in the kinetic energy for the last and first points and dividing it by the difference in frequencies, find the slope of the electron kinetic energy vs. frequency curve. Compare your answer with the value of h.

d. Check the validity of using the small momentum approximation to the energy. In each case, calculate the electron kinetic energy using the non-relativistic approximation, then observe in what significant figure the more accurate result differs. Use the values of the momentum found in part a.

Results such as these form the basis for the famous Einstein photoelectric equation $K = h\nu - \phi$. The importance of the calculation lies in the treatment of the photon as a particle with energy $h\nu$ and momentum $h\nu/c$.

Notice the role played by the recoil nucleus. It carries away very little of the original energy of the photon but the magnitude of its momentum is nearly the same as the magnitude of the electron momentum. The ratio of the kinetic energies K_e/K_M is nearly the same as the mass ratio M/m and the small difference in the momentum accounts for the photon momentum. The heavy particle recoils in a direction opposite to the direction in which the electron is ejected.

Nearly the same results are obtained if the electron is ejected in other directions. The ejected electrons have nearly the same kinetic energy, regardless of the direction in which they are ejected. This is because pc is much larger than $h\nu$. In Eqn. 25-12, the term $(pc)^2$ dominates the other two terms and, as a consequence, the momenta of the electron and recoil nucleus have nearly the same magnitude, regardless of the value of θ_1.

Problem 2. Light of frequency 4×10^{15} Hz is incident on a potassium atom. For each of the following directions of electron emission, find the momentum and kinetic energy of the electron and of the recoiling potassium atom. Again take $\phi = 6.92 \times 10^{-19}$ J. Compare the answers to those of problem 1 to see the extent to which the momentum and energy of the ejected electron are different for different angles of emission.

a. $\theta_1 = 0$.

b. $\theta_1 = 60^{\circ}$.

c. $\theta_1 = 90^{\circ}$.

In the usual photoelectric effect experiment, not all the ejected electrons have the same kinetic energy after emission. The variation in energy is clearly not due to a variation in the direction of emission but rather it is due to the variation in the initial energy of the electrons. Some electrons are more tightly bound than others and, for them, a larger fraction of the photon energy is required to free them. The most loosely bound electrons are ejected with the greatest kinetic energy.

The photoelectric effect can be used to determine the energies of electrons in materials. It is a very delicate experiment, however, and data are difficult to interpret since the results are quite sensitive to the condition of the material surface.

In one regard, the above problems are not typical. Although photoelectric experiments are performed on single atoms in a gas, the usual target electron is in solid material and the mass of the recoil matter is much larger than that of a single atom. The recoil matter carries away a much smaller fraction of the original photon energy.

To demonstrate the influence of the recoil mass on the photoelectric effect, we consider a mass which is smaller than that of a potassium atom and compare the results with those of problem 1.

Problem 3. A hydrogen atom consists of a proton (mass M=1.67265×10^{-27} kg) and an electron with a binding energy of about 2.18×10^{-18} J. The electron absorbs a photon and is ejected in the direction opposite to that of the incident photon. Assume both the proton and the electron are initially at rest.

a. For each of the following frequencies of light, find the momentum and kinetic energy of both the ejected electron and the proton, after the interaction. Obtain five significant figures for the root pc.

i. $\nu=4\times10^{15}$ Hz.

ii. $\nu=6\times10^{15}$ Hz.

iii. $\nu=8\times10^{15}$ Hz.

b. For each frequency, calculate $h\nu-\phi$ and compare the result with the kinetic energy of the ejected electron.

c. Find the slope of the electron kinetic energy vs. frequency curve. Do this by finding the difference in the kinetic energy for the last and first points and dividing it by the difference in frequencies. Compare your answer with the value of h.

Now the recoil mass carries off a larger fraction of the original photon energy than was the case for the potassium atom and the slope of the electron kinetic energy vs. frequency curve differs slightly more from h.

The photoelectric effect can occur for any frequency of incident light. If $h\nu$ is comparable to or greater than the electron rest energy, then the relativistically correct relationship between energy and momentum must be used. The electron momentum is found by seeking the root of

$$\left[(mc^2)^2+(pc)^2\right]^{\frac{1}{2}} + \left[(Mc^2)^2+(Pc)^2\right]^{\frac{1}{2}} - mc^2 - Mc^2 - h\nu + \phi \qquad (25\text{-}14)$$

where $(Pc)^2$ is again given by Eqn. 25-12.

Problem 4. The electron in a hydrogen atom absorbs a photon and is ejected in the direction opposite to that of the incoming photon. Assume both the proton and the

502

electron are initially at rest. Take $\phi=2.18\times10^{-18}$ J. This term does not play an important role for the photon energies to be considered.

a. Write instructions for the binary search program so it can be used to find the roots of expression 25-14.

b. For each of the following frequencies of light, use the program to find the momentum and kinetic energy of both the ejected electron and the proton, after the interaction. Obtain five significant figures for the root.

   i. $\nu=5\times10^{19}$ Hz.
   ii. $\nu=1\times10^{20}$ Hz.
   iii. $\nu=1.5\times10^{20}$ Hz.
   iv. $\nu=2\times10^{20}$ Hz.

c. For each frequency, calculate $h\nu-\phi$ and compare the result with the kinetic energy of the ejected electron. Does the electron kinetic energy vary linearly with the frequency of the incident light?

25.3   The Compton Effect

A photon need not be absorbed by an electron when they interact. Instead, it may be scattered and move off in a direction which is different from the direction of incidence. If the photon energy is sufficiently great, the frequency of the scattered light is significantly less than that of the incident light and, furthermore, the frequency is different for different directions of emergence of the light. This phenomenon is known as the Compton effect.

Qualitatively, the change in frequency is easy to understand. The photon loses energy to the electron and, because the photon energy and wave frequency are proportional to each other, the frequency is reduced. Since the electron and photon momentum, after the collision, must sum to the total momentum before the collision, the momentum transferred during the collision is different for different scattering angles. This leads to the dependence of the energy transfer on the scattering angle.

The fundamental problem is to find the wavelength and frequency of the scattered light as a function of scattering angle. At first, we consider photon energies which are much greater than the binding energy of the electron. The binding energy can then be neglected and the electron can be considered to be free. We also take the electron to be initially at rest.

The technique used is similar to that used to analyze the photoelectric effect. The photon is a particle with energy $h\nu$ and momentum $h\nu/c$. It is incident in the positive x direction on an electron at rest. We suppose that, after the collision, the photon moves off in the x,y plane at an angle θ_1 above the x axis and that the electron moves off in the x,y plane at an angle θ_2 below the x axis. After the interaction, the photon has energy $h\nu'$ and momentum $h\nu'/c$, while the electron has momentum p.

The conservation laws for energy and momentum yield

$$h\nu + mc^2 = h\nu' + \left[(mc^2)^2 + (pc)^2\right]^{\frac{1}{2}}, \qquad (25\text{-}15)$$

$$h\nu = h\nu'\cos\theta_1 + pc\,\cos\theta_2, \qquad (25\text{-}16)$$

and
$$0 = h\nu'\sin\theta_1 - pc\,\sin\theta_2. \qquad (25\text{-}17)$$

We suppose the photon is observed and the angle θ_1 is measured. The angle θ_2 can be eliminated, as usual, by solving Eqn. 25-16 for $\cos\theta_2$, Eqn. 25-17 for $\sin\theta_2$, then squaring and adding. The result is solved for $(pc)^2$:

$$(pc)^2 = (h\nu)^2 - 2(h\nu)(h\nu')\cos\theta_1 + (h\nu')^2. \qquad (25\text{-}18)$$

This is substituted into Eqn. 25-15 to find that

$$h\nu' - h\nu - mc^2 + \left[(mc^2)^2 + (h\nu)^2 - 2(h\nu)(h\nu')\cos\theta_1 + (h\nu')^2\right]^{\frac{1}{2}} = 0. \qquad (25\text{-}19)$$

Given the frequency ν and the scattering angle θ_1, one of the root finding programs can be used to find ν'. Once ν' is known, Eqn. 25-18 can be used to find the magnitude of the electron momentum. Finally, Eqns. 25-16 and 25-17 can be used to find the direction in which the electron travels after the interaction.

We describe the instructions for the binary search program. Similar instructions can be used with the method of uniform intervals. It is convenient to use $h\nu'$ as the variable and two values which straddle the root are placed in X8 and X9, respectively. Line 510 becomes

```
(hν)↑2+X8*(X8-2*(hν)*COS(θ₁))→X10
```

```
X8-(hν)-(mc²)+SQRT((mc²)↑2+X10)→X2
```

Here $(pc)^2$ is first evaluated and stored in X10, then the left side of Eqn. 25-19 is evaluated. Similar instructions are used at lines 520 and 600. There, either X9 or X4 replaces X8 and the result is placed in either X3 or X5, as appropriate. When the root has been found, it must be divided by h to find the frequency ν'. If your machine has sufficient memory locations available, it is worthwhile to store values of $h\nu$ and θ_1, then recall them as needed. This saves rewriting instructions when the problems call for different values of these parameters.

Problem 1.   Light with wavelength 4.5×10^{-12} m is incident on a free electron at rest.

a.   For each of the following scattering angles, use a root finding program to evaluate the frequency and wavelength of the scattered light and the kinetic energy and magnitude of the momentum of the scattered electron. Obtain 5 significant figures for the root.

   i. $\theta_1 = 0$.
   ii. $\theta_1 = 40°$.
   iii. $\theta_1 = 90°$.
   iv. $\theta_1 = 120°$.
   v. $\theta_1 = 180°$.

b.   For each of the above scattering angles, find the change $\Delta\lambda = \lambda' - \lambda$ in the wavelength on scattering.

c.   After the expression for $(pc)^2$, given by Eqn. 25-18, has been substituted into Eqn. 25-15, the resulting equation can be solved algebraically for $\lambda' - \lambda$. Do this and show that

$$\lambda' - \lambda = \frac{hc}{mc^2}(1 - \cos\theta_1).$$

d.   As a check on the results obtained numerically, evaluate $\Delta\lambda$, using the expression found in part c, for each of the scattering angles of part a.

   When $\theta_1 = 0$, no change in the wavelength occurs and the electron remains at rest. The change in wavelength increases as θ_1 increases and it is the greatest for $\theta_1 = 180°$. For this scattering angle, the energy and momentum transferred to the electron are greater than for any other scattering angle.

   The change in the wavelength is independent of the frequency of the incident light. This conclusion is substantiated by the equation derived in part c of the above problem, but let us verify it using numerical techniques.

Problem 2. Monochromatic light is incident on a free electron at rest and light scattered through 40° is observed.

a. For each of the following wavelengths of the incident light, use a root finding program to calculate the frequency and wavelength of the scattered light and the kinetic energy and magnitude of the momentum for the scattered electron. Obtain eight significant figures for the root.

   i. $\lambda = 4.5 \times 10^{-8}$ m. (There may be some round off error.)

   ii. $\lambda = 4.5 \times 10^{-10}$ m.

   iii. $\lambda = 4.5 \times 10^{-14}$ m.

b. For each of the wavelengths given above, calculate the change $\Delta\lambda = \lambda' - \lambda$ in the wavelength during scattering and verify that it is the same for all wavelengths of the incident light.

   Since the wavelength shift is the same for all wavelengths, the fractional shift is small if the wavelength of the incident light is large and large if the wavelength of the incident light is small. The shift is an appreciable fraction of the original wavelength if that wavelength is on the order of $hc/(mc^2) = 2.4 \times 10^{-12}$ m or shorter. This corresponds to a frequency of 1.2×10^{20} Hz or greater. These waves are in the regions of the electromagnetic spectrum which are known as the x-ray and gamma ray regions.

   For a Compton effect experiment, the intensity of scattered light for a selected scattering angle is sometimes plotted as a function of wavelength. These plots show spikes at the wavelengths λ' calculated using the numerical or algebraic techniques of problem 1. The spikes are narrow but they do have some width, indicating that light with a range of wavelengths emerges at any selected angle.

   The wavelength of the scattered light has a value which is different from that predicted by Eqn. 25-19 because the electron is not initially at rest. If the electron is initially moving at an angle α with the x axis and if it has momentum with magnitude p_0, then the conservation laws become

$$h\nu + p_0 c \cos\alpha = h\nu' \cos\theta_1 + pc \cos\theta_2, \qquad (25\text{-}20)$$

506

$$p_0 c \sin\alpha = h\nu' \sin\theta_1 - pc \sin\theta_2, \qquad (25\text{-}21)$$

and
$$h\nu + \left[(mc^2)^2 + (p_0 c)^2\right]^{\frac{1}{2}} = h\nu' + \left[(mc^2)^2 + (pc)^2\right]^{\frac{1}{2}}. \qquad (25\text{-}22)$$

Eqns. 25-20 and 25-21 can be solved for $(pc)^2$, this time in terms of p_0, α, ν, ν', and θ_1. The result is

$$(pc)^2 = (h\nu')\left[h\nu' - 2h\nu \cos\theta_1 - 2p_0 c \cos(\alpha-\theta_1)\right]$$
$$+ (h\nu)\left[h\nu + 2p_0 c \cos\alpha\right] + (p_0 c)^2. \qquad (25\text{-}23)$$

This expression is substituted into Eqn. 25-22 and the result is solved for $h\nu'$.

If the binary search program of Fig. 2-3 is used, two values of $h\nu'$ which straddle the root are placed in X8 and X9 respectively and the instruction at line 510 is changed to read

$$\text{X8*(X8-2*(h}\nu)\text{COS}(\theta_1)\text{-2*(p}_0\text{c)*COS}(\alpha-\theta_1))\text{+(h}\nu)\text{*((h}\nu)\text{+2*(p}_0\text{c)}$$

$$\text{*COS}(\alpha))\text{+(p}_0\text{c)}\uparrow 2 \rightarrow \text{X10}$$

$$\text{X8-(h}\nu)\text{+SQRT((mc}^2)\uparrow 2\text{+X10)-SQRT((mc}^2)\uparrow 2\text{+(p}_0\text{c)}\uparrow 2) \rightarrow \text{X2}$$

Similar instructions are used at lines 520 and 600, with appropriate replacements for X8 and X2. Finally, the root, when found, is divided by h to produce the value ν' of the frequency. If storage space is available, values of $h\nu$, $p_0 c$, θ_1, and α should be stored at the beginning of the program and recalled as needed.

Problem 3. Light of wavelength 4.5×10^{-12} m is incident along the x axis on a moving electron and the light which is scattered through 40° is observed. The electron initially has momentum $p_0 = 1 \times 10^{-24}$ kg·m/s and moves along the line at the angle α with the positive x direction. For each of the following values of α, use a root finding program to solve for the frequency of the scattered light, then find the change in wavelength. Obtain five significant figures for the root. Also find the magnitude of the momentum and the kinetic energy of the scattered electron.

a. $\alpha=0$.

b. $\alpha=40^\circ$.

c. $\alpha=90^\circ$.

d. $\alpha = 120^{\circ}$.

e. $\alpha = 180^{\circ}$.

The shift in wavelength is less when the electron is initially moving toward the incoming photon than when it is moving away. Larger wavelength shifts are obtained if the initial electron momentum is greater in magnitude.

In any material, electrons are present with a large variety of speeds, moving in all directions. For a given direction of incidence and a given scattering angle, the wavelength of the scattered light is distributed over a range of values. It is not monochromatic as it would be if the electrons were initially at rest. The experimentally measured distribution of wavelengths has been used to study the momenta of the electrons in materials.

Problems involving an electron initially in motion can be worked using the rest frame of the electron, then applying the Lorentz transformation. The following problem is presented as an example of this technique.

Problem 4.   Light with wavelength 4.5×10^{-12} m is incident along the x axis on a free electron at rest. A photon is scattered through 40°. This is the same situation as was described in problem 1. Use your answers to that problem.

a.   Find the speed of a frame in which the electron initially has momentum 1×10^{-22} kg·m/s, in the negative x direction.

b.   Use the Lorentz transformation to carry out the following calculations.

   i.   Find the frequency and wavelength of the incident light, as measured in the frame of part a.

   ii.   For the frame of part a, find the photon energy and components of the photon momentum, after scattering. Then find the frequency and wavelength of the scattered light. To do this, transform the results of problem 1.

   iii. Find the direction of motion of the scattered photon, as measured in the frame of part a.

c.   Now solve the conservation of energy and momentum equations for the frequency and wavelength of the scattered light, in the frame of part a. The wavelength of the incident photon and the scattering angle have the values found in part b. Compare your answer with that found using the Lorentz transformation.

For a given scattering angle, a large portion of the scattered light has a wavelength which is not measurably different from that of the incident light. These photons have been scattered from atoms, which are much more massive than electrons. In the next problem, scattering from a proton is considered as an example to demonstrate that the wavelength change is less for scattering from more massive particles.

Problem 5. Light of wavelength 4.5×10^{-12} m is incident on a proton at rest. Light scattered through 90° is observed. Use a root finding program to find the frequency and wavelength of the scattered light as well as the kinetic energy and magnitude of the momentum of the scattered proton. Compare the results with those for scattering from an electron. In particular, calculate the ratio of the wavelength shift for proton scattering to the wavelength shift for electron scattering and compare the result with the ratio of the electron mass to the proton mass.

25.4 Electron-Positron Annihilation

A positron is the antiparticle to the electron. An electron and a positron have the same mass and the same magnitude of charge. Positrons, however, are positively charged while electrons are negatively charged.

Upon interacting, an electron and a positron can annihilate each other. Both particles disappear and other particles appear to carry the original energy and momentum. If the electron and positron have small kinetic energies, the most likely particles produced by the annihilation are photons. We consider such events.

Mutual annihilation of a free electron and a free positron produces at least two photons. That a single photon cannot be produced is most easily seen by considering the interaction in the center of mass frame. For that frame of reference, the total momentum before the interaction is zero and it must also be zero after the interaction. Since a single photon must carry some momentum, this result violates the conservation laws. Furthermore, if the interaction must produce at least two photons in the center of mass frame, then it must produce at least two photons in every frame.

If annihilation takes place in the vicinity of a third particle, such as a nucleus, the third particle can carry some of the original momentum and, in this case, a single photon can be the result. Some annihilation interactions produce three or more photons. We consider the case of two photons. Again the photons are treated as particles and the conservation laws are invoked.

Both the electron and the positron are assumed to be moving along the x axis, the electron with momentum p_- and the positron with momentum p_+. The x component of the total momentum is

$$p = p_- + p_+ \qquad\qquad (25-24)$$

and the other components vanish. The total energy is given by

$$E = \left[(mc^2)^2 + (p_-c)^2\right]^{\frac{1}{2}} + \left[(mc^2)^2 + (p_+c)^2\right]^{\frac{1}{2}} \qquad\qquad (25-25)$$

where m is the mass of either particle.

After the interaction, one photon has energy $h\nu_1$ and moves along the line at the angle θ_1 above the x axis, while the other photon has energy $h\nu_2$ and moves along the line at the angle θ_2 below the x axis. The conservation laws produce the equations

$$E = h\nu_1 + h\nu_2, \qquad\qquad (25-26)$$

$$pc = h\nu_1 \cos\theta_1 + h\nu_2 \cos\theta_2, \qquad\qquad (25-27)$$

and $$0 = h\nu_1 \sin\theta_1 - h\nu_2 \sin\theta_2. \qquad\qquad (25-28)$$

Notice that the rest energies of the electron and positron are included as part of the original total energy. These particles disappear during the interaction and their rest energies become part of the final photon energy. Mass is not conserved during the event. Both the electron and positron have non-vanishing mass while the photon is massless.

For the following problems, we assume p and E are known or can be calculated easily from given quantities. The angle at which one of the photons leaves the interaction is also given and the frequencies of the photon waves as well as the direction of travel of the second photon are to be computed.

Solve Eqn. 25-27 for $\cos\theta_2$, solve Eqn. 25-28 for $\sin\theta_2$, then square and add the resulting expressions. Finally solve for $h\nu_2$:

$$hv_2 = \left[(pc)^2 - (hv_1)(2pc\ cos\theta_1 - hv_1)\right]^{\frac{1}{2}}. \tag{25-29}$$

This result is substituted into Eqn. 25-26 and the root of $E-hv_1-hv_2$, a function of hv_1, is sought.

The energy of the second photon can be found using Eqn. 25-26 and the direction of of travel of the second photon can be found using

$$\theta_2 = \tan^{-1} \frac{hv_1\ sin\theta_1}{pc-hv_1\ cos\theta_1}. \tag{25-30}$$

This equation results when $sin\theta_2$, from Eqn. 25-28, is divided by $cos\theta_2$, from Eqn. 25-27.

It is convenient to use the product hv_1 as the variable in the root finding process. For the binary search program of Fig. 2-3, two values of hv_1 which straddle the root are placed in X8 and X9 respectively. The instructions at line 510 are

```
SQRT((pc)↑2-X8*(2*(pc)*COS(θ₁)-X8))→X10
X8+X10-(E)→X2
```

Similar instructions are programmed at lines 520 and 600 with appropriate replacements for X8 and X2. When the root has been found, it is divided by h to find the frequency v_1. Then v_2 and θ_2 can be found, if required. There are two angles which obey Eqn. 25-30 and care must be taken to select the one which satisfies Eqns. 25-27 and 25-28 separately.

We first consider a positron incident on an electron at rest and calculate the frequencies of the photon waves for photons emitted along the positron line of motion. If one photon is emitted in the same direction as the incident positron, the conservation laws dictate that the other photon is emitted along the same line.

Problem 1. A positron, moving in the positive x direction, is incident on an electron at rest. Two photons are emitted and one of them travels in the positive x direction.
a.   For each of the following values of the positron momentum, use a root finding
     program to calculate the energies of the photons and the frequencies of their waves.

i. $p_+ = 4 \times 10^{-20}$ kg·m/s.

ii. $p_+ = 4 \times 10^{-21}$ kg·m/s.

iii. $p_+ = 4 \times 10^{-22}$ kg·m/s.

iv. $p_+ = 4 \times 10^{-23}$ kg·m/s.

v. $p_+ = 4 \times 10^{-24}$ kg·m/s.

b. As a check on the program, verify in each case that the photon momenta sum to give the original positron momentum and that the photon energies sum to give the original energy, the sum of the positron rest energy, the electron rest energy, and the positron kinetic energy. Be sure to use the correct signs for the photon momenta.

c. For the smallest positron momentum considered, the ratio of the photon energies is about 1.01, while for the largest considered, it is about 293. Use the conservation laws to explain qualitatively why the ratio increases with positron momentum.

The photon energies are dependent on the angle of emission. For the same positron momentum, photons which are emitted in different directions have different energies. These photons are produced by different annihilation events, of course.

Problem 2.   An experiment consists of the annihilation of an electron at rest by a positron moving in the positive x direction with momentum $p_+ = 4 \times 10^{-22}$ kg·m/s. The experiment is repeated three times, and each time two photons are emitted, but in different directions for different experiments.

a. For each experiment, the angle θ_1 between the direction of travel for one of the photons and the x axis is given below. In each case, find the energies of the photons and the frequencies of their waves. Also find the direction of travel of the second photon.

i. $\theta_1 = 40^\circ$.

ii. $\theta_1 = 90^\circ$.

iii. $\theta_1 = 120^\circ$.

b. If a photon were emitted at $\theta_1 = 180^\circ$, what would the energies of the two photons be? Use the results of problem 1 to answer this question.

In the rest frame of the electron-positron system, the total momentum vanishes and, according to the conservation laws, $h\nu_1 = h\nu_2 = E/2$, where E is the total energy in that frame. The photon momenta are equal in magnitude but in opposite directions.

Problem 3.   A positron, moving in the positive x direction with momentum 4×10^{-22} kg·m/s, is incident on an electron at rest. Two photons are emitted, one in the positive x direction and the other in the negative x direction.

a.   Find the velocity of a reference frame in which the total momentum vanishes.

b.   Use the Lorentz transformation to find the total energy in the frame of part a. Don't forget the electron rest energy.

c.   Use $h\nu_1=h\nu_2=E/2$ and $p_1=h\nu_1/c$, $p_2=-h\nu_2/c$ to find the energy and momentum of each of the photons, as measured in the frame of part a.

d.   Use the Lorentz transformation to find the energy of each photon and the frequency of each of the photon waves, as measured in the original frame of reference. Compare the results with the answers obtained to the appropriate parts of problem 1.

    Positron annihilation is used to measure the momentum and energy of electrons in atoms and in solids. If the momentum of the positron is known and properties of the emitted photons are measured, it is possible to calculate the energy and momentum of the electron annihilated by the positron.

    In these experiments, it is usual to measure the angles at which the photons are emitted rather than their energies. This is because the angles can be measured with higher precision than the energies.

    We take a simple case to demonstrate how the procedure works. Suppose the electron is moving along the x axis. Then Eqns. 25-26, 25-27, and 25-28 are valid. Solve Eqn. 25-28 for $h\nu_1$:

$$h\nu_1 = h\nu_2 \; \sin\theta_2/\sin\theta_1. \tag{25-31}$$

Substitute the result into Eqn. 25-27 and solve for $h\nu_2$:

$$h\nu_2 = pc\,\frac{\sin\theta_1}{\sin(\theta_1+\theta_2)}. \tag{25-32}$$

The trigonometric identity $\sin(\theta_1+\theta_2)=\sin\theta_1\cos\theta_2+\cos\theta_1\sin\theta_2$ was used. Substitution of these expressions for $h\nu_1$ and $h\nu_2$ into Eqn. 25-26 yields

$$E = pc\,\frac{\sin\theta_1 + \sin\theta_2}{\sin(\theta_1+\theta_2)}. \tag{25-33}$$

Since $E = \left[(mc^2)^2 + (p_- c)^2\right]^{\frac{1}{2}} + \left[(mc^2)^2 + (p_+ c)^2\right]^{\frac{1}{2}}$ and $p = p_- + p_+$, it follows that

$$\left[(mc^2)^2 + (p_- c)^2\right]^{\frac{1}{2}} + \left[(mc^2)^2 + (p_+ c)^2\right]^{\frac{1}{2}} = (p_- + p_+)c\,\frac{\sin\theta_1 + \sin\theta_2}{\sin(\theta_1 + \theta_2)}\,. \qquad (25\text{-}34)$$

This equation is to be solved for the electron momentum, given p_+, θ_1, and θ_2.

Problem 4.   A positron moves in the positive x direction with momentum $p_+ = 4.2 \times 10^{-23}$ kg·m/s. It annihilates an electron, also moving along the x axis. One photon is emitted at an angle of 90° above the x axis and another is emitted at an angle of 73° below the x axis.

a.   Modify a root finding program to solve Eqn. 25-34 for the electron momentum p_-.
    Use $p_- c$ as the variable.

b.   Run the program and solve for p_-.

c.   Find the energies of the two emitted photons and the frequencies of their waves.

    In actual practice, the momentum of the positron just before annihilation is not the momentum with which it is fired into the material. Before annihilating an electron, the positron is usually slowed by collisions with atoms of the material. It may then be captured by an electron and the two particles may go into orbit about each other, held by their mutual electrical attraction, which acts as the source of a centripetal force. During this process, the particles may lose energy via radiation. Finally, they annihilate each other and emit more photons. The experimenter must be careful to measure the angles of emission for photons from the annihilation rather than those emitted by the electron-positron system before annihilation.

    The process which is the reverse to pair annihilation, called pair production, also occurs in nature. In this process, one or more photons disappear and, in their place, a particle and its antiparticle appear. It is clear that the total photon energy must at least be equal to the twice the rest energy of either of the particles in the pair. We shall not study this process here.

Chapter 26

MATTER WAVES

In this chapter, the calculator is used to plot probability distributions for particles and to calculate the probability of obtaining selected values of the particle coordinate, energy, and momentum when these quantities are measured. A detailed demonstration of the uncertainty principle is given. These plotting exercises and calculations are designed to help produce a better understanding of the quantum mechanical behavior of particles. They supplement Chapter 50 of PHYSICS, Chapter 43 of FUNDAMENTALS OF PHYSICS, Chapter 6 of RELATIVITY AND EARLY QUANTUM THEORY, and the quantum mechanics sections of other texts.

26.1 <u>Particle</u> <u>Wave</u> <u>Functions</u>

Just as for photons, there is a probability density amplitude wave associated with other particles. Such a wave is associated with an electron, for example.

Particle waves are usually denoted by the symbol Ψ and are functions of position in space and the time. We consider waves which are functions of x and t only and write $\Psi(x,t)$ for the particle wave function.

The physical interpretation of a particle wave function is the same as the interpretation of the wave function for light. In the one dimensional case,

$$dP = \left| \Psi(x,t) \right|^2 dx \qquad\qquad (26\text{-}1)$$

gives the probability that, at time t, the particle is between x and x+dx.

Just as for photons, it is impossible to control an electron in such a way that it arrives at a specified region of space at a specified time. The outcomes of many experiments, each with the same initial conditions and each with the same forces acting, might all be quite different. All that can be predicted is the probability of a specified outcome.

If a great many position measurements are performed, each starting with the same initial conditions, then $|\Psi|^2$ dx gives the fraction of these which give the result that the particle is between x and x+dx.

Each time the experiment is repeated, the particle must be placed in the original state. A measurement of position usually alters the state and the wave function of the particle. The probability being discussed is the probability that the particle is in a certain region, provided it is in the specified state with the specified wave function. Once the measurement is taken, it no longer has that wave function and the probability is altered. To obtain a second measurement, with the particle in the original state, the experiment must be started over.

For free electrons, not subject to forces, the wave function oscillates in both space and time. Just as for photons, the wave frequency is E/h where E is the particle energy and the wavelength is h/p where p is the particle momentum. The mathematical description of these waves is somewhat different from the description of photon waves and we defer quantitative discussion of them until later.

Diffraction and interference effects occur for particles with mass as well as for light. If electrons are fired at a pair of slits and then detected on the other side, the intensity pattern is very nearly the same as the intensity pattern for light. The electrons arrive randomly at the detectors but they are more likely to strike the screen at places where the square of the wave amplitude is high and less likely to strike at places where the square of the wave amplitude is low. The pattern might look like the distribution you plotted in problem 1 of section 24.1. In practice, this experiment is extremely difficult to perform since, for typical electrons in the laboratory, the slits must have widths on the order of tens of angstroms to see an effect.

Mathematically, interference effects are produced by summing the probability density amplitude waves from the slits. These waves may be out of phase at some points on the screen. They then interfere destructively there and the probability of finding the electron at those places is small. At other places, the waves may interfere constructively and the probability of finding the electron at those places is large.

Cutting another slit in the barrier changes the distribution of probability at the screen since this means other waves are added to those from the previously existing slits. This action causes the probability at some points on the screen to decrease and the probability at other points to increase.

When the particle is not free, the wave function and the probability pattern may be quite different from those studied in Chapter 22. In the remainder of this section, some probability patterns are examined for several physical situations. These are all associated with non-relativisitic particles. The ideas are the same for relativistic particles but the mathematics is somewhat more complicated.

As an example of a matter wave, we first consider the one dimensional case of a particle confined by means of rigid walls to the region between x=0 and x=L. The particle is said to be in a box. The wave function has nodes at the walls and is zero beyond them. It is a standing wave of the form

$$\Psi_n = (2/L)^{\frac{1}{2}} \sin(\frac{n\pi x}{L}) \, f_n(t) \qquad (26\text{-}2)$$

where n is one of the integers 1, 2, 3, ... and $f_n(t)$ is a function of time. Verify that Ψ_n=0 at x=0 and x=L. The argument of the sine function is in radians.

The function f_n is an oscillatory function with frequency ν=E/h, where E is the energy of the particle. This function has magnitude 1 and, when the expression for the probability density is developed, f_n does not appear. We leave it unspecified for the moment, but will return to it later.

When the particle wave function is given by Eqn. 26-2, the energy of the particle is

$$E_n = \frac{n^2 h^2}{8mL^2} \, , \qquad (26\text{-}3)$$

where m is the mass of the particle and h is Planck's constant. The energy must have one of these values. The particle cannot have any other energy and be confined to the box.

That there are discrete energy levels for the particle in a box is one example of a general principle. If a particle is bound, its energy is quantized. That means it must have a value which is one of a set of discrete values. Any two allowed energy values are separated by a range of unallowed values.

On the other hand, if the particle is not bound, the allowed energy values form a continuous set. The energy is not quantized.

For the particle in a box, the smallest energy is $E_1 = h^2/(8mL^2)$. The particle cannot be confined to the box if it has less energy. If the particle energy is to be increased, energy must be added in the proper increments so that the final energy satisfies Eqn. 26-3.

The probability density for a particle with the wave function Ψ_n is

$$|\Psi_n|^2 = (2/L)\ \sin^2(\frac{n\pi x}{L}).\qquad\qquad (26\text{-}4)$$

The magnitude squared of the function $f_n(t)$ is unity and the probability density does not depend on the time. It does, however, depend on x. There is a higher probability that the particle can be found in some regions of the box than in other regions.

Problem 1.   An electron is confined to the region of the x axis between x=0 and $x=1.6\times10^{-8}$ m. Suppose it is in the state with the lowest energy.

a.   Calculate the value of the electron energy.

b.   Use the program of Fig. 2-4 to plot the probability density as a function of x. Use 1×10^{-9} m intervals.

c.   On the graph, locate the place of maximum probability density.

For states with higher energy, nodes appear in the probability density function. There are places where there is zero probability of finding the particle.

Problem 2.   An electron is confined to the region of the x axis between x=0 and $x=1.6\times10^{-8}$ m. It is in the state with n=5.

a.   Calculate the value of its energy.

b.   Use the program of Fig. 2-4 to plot the probability density as a function of x. Use 5×10^{-10} m intervals.

c.   On the graph, locate the places of maximum probability density and the places of zero probability density.

The probability of finding the particle in a finite segment of the x axis is calculated by integrating the probability density over that segment. For example, the probability of finding the particle between $x=x_1$ and $x=x_2$ is given by

518

$$P = \int_{x_1}^{x_2} |\Psi|^2 \, dx. \tag{26-5}$$

For the particle in a box, this is

$$P = (2/L)\int_{x_1}^{x_2} \sin^2(\frac{n\pi x}{L}) \, dx, \tag{26-6}$$

provided x_1 and x_2 are within the box.

The probability of finding the particle somewhere in the box must be unity. In fact, the constant $(2/L)^{\frac{1}{2}}$ in the wave function was chosen so that

$$(2/L)\int_{0}^{L} \sin^2(\frac{n\pi x}{L}) \, dx = 1. \tag{26-7}$$

This integral sums the probabilities for all the segments of the x axis within the box.

The integration program of Fig. 8-1 can be used to evaluate Eqn. 26-5. The lower limit x_1 is entered into X1, the upper limit x_2 is entered into X2, and the number of segments N is entered into X3. $|\Psi|^2$ is evaluated at lines 130, 170, and 180. For the particle in a box, the instructions at line 130 are

SIN((nπ/L)*X1)↑2*(2/L)→X5

At lines 170 and 180, the instruction is the same, except X5 is replaced by X9.

Problem 3. An electron is confined to the region of the x axis between x=0 and x=1.6×10^{-8} m. It is in the state with n=4. For each of the following intervals, use the integration program of Fig. 8-1, with N=30, to find the probability of finding the electron in that interval.

a. $x_1=0$, $x_2=L/2$.
b. $x_1=L/2$, $x_2=L$.
c. $x_1=0$, $x_2=L/3$.
d. $x_1=L/3$, $x_2=2L/3$.
e. $x_1=2L/3$, $x_2=L$.

519

The probabilities found in parts a and b should sum to 1, as should those found in parts c, d, and e.

The probability of finding the electron in a particular segment of the x axis may be different for different values of the electron energy.

Problem 4.   An electron is confined to the region of the x axis between x=0 and x=1.6×10^{-8} m. For each of the following states, use the program of Fig. 8-1 to find the probability the electron is found between x=3×10^{-9} m and x=5×10^{-9} m.

a.   n=1. Use N=20.

b.   n=2. Use N=20.

c.   n=4. Use N=20.

d.   n=20. Use N=30.

e.   n=50. Use N=50.

For low energies, the probability is strongly dependent on the value of the energy, while, for high energy, it is not. For high energy, the segment contains many spatial oscillations of the wave function and nearly the same value of the probability is obtained for all segments of equal length. This value is the ratio of the segment length to the box length. Nearly the same result is obtained no matter what the energy, as long as it is great enough that many oscillations fit into the segment.

Problem 5.   An electron, confined to a box of width 1.6×10^{-8} m, is in the state with n=4. A photon is emitted and the electron changes state to the one with n=3.

a.   Use the program of Fig. 2-4 to plot, on the same graph, the probability density before and after the photon emission. Plot points every 5×10^{-10} m.

b.   What is the energy of the photon and the frequency of the photon wave? Assume the box is sufficiently massive that the energy it receives is negligible.

c.   What is the probability of finding the electron in the segment from x=7×10^{-9} m to x=9×10^{-9} m, if it is detected before the photon is emitted? What is the probability of finding it in the same segment if it is detected after the photon is emitted? Use the program of Fig. 8-1 with N=20.

When the energy of the particle is increased or decreased, as in the above problem, the probability density function changes. The probability of finding the particle in any particular segment of the x axis is different after the photon is emitted than it was before.

In classical mechanics, forces determine the acceleration of the particle and, in conjunction with initial conditions, determine the position of the particle as a function of time. This description of particle mechanics is valid for large objects. At the atomic level, however, this description cannot be valid. Where the particle goes is a matter of chance and the chances are determined by the wave function associated with the particle.

What then is the role of force in the quantum description of nature? The answer is that forces determine the probability density amplitude waves. Two electrons which are subject to different forces have different waves associated with them. In the modern quantum theory, the wave function is determined by the potential energy function for the particle and this function is, in turn, determined by the forces which act.

We have examined the wave function and probability density for a particle acted on by strong forces exerted by the walls of a box to keep the particle inside. As an example of the wave function produced by another force, we now consider a particle acted on by a spring-like force. The particle is a harmonic oscillator.

For a harmonic oscillator, the force is given by $F=-kx$ and the potential energy by $V=\frac{1}{2}kx^2$, where k is the spring constant. It is usual to write this $V=\frac{1}{2}m\omega_0^2x^2$, where $\omega_0=(k/m)^{\frac{1}{2}}$. This is the natural angular frequency of the oscillator and should not be confused with the angular frequency of the particle wave.

Although atoms and atomic particles are never attached to springs as such, some forces which do act on them are proportional to their displacements and act like spring forces. When an atom in a crystalline solid, for example, is displaced slightly from its equilibrium position, a restoring force proportional to the displacement acts on it and it oscillates about its equilibrium position. The force arises from the electrostatic forces exerted by neighboring atoms in the crystal.

For an oscillator, the particle is bound and the energy is again quantized. The particle cannot have any energy but, rather, its energy must have one of the values

$$E_n = (\hbar\omega_0/2\pi)(n+\tfrac{1}{2}), \quad n=0, 1, 2, \ldots \tag{26-8}$$

The allowed energy levels are uniformly spaced with a separation of $\hbar\omega_0/2\pi$. If it is desired to increase the energy of the oscillator, energy must be added in integer multiples of $\hbar\omega_0/2\pi$. The particle cannot accept other increments of energy.

The energy, as given by Eqn. 26-8, is measured relative to the energy for a situation in which the force constant and the natural frequency are zero. Even in the lowest energy state, for which $n=0$, the particle has some energy relative to the reference situation. This energy, called the zero point energy, is given by $E_0 = \tfrac{1}{2}\hbar\omega_0/2\pi$. That the oscillator has a non-vanishing zero point energy is an indication that the particle cannot be acted on by a spring-like force without some motion taking place.

If the laws of classical mechanics were valid, the particle could be placed at its equilibrium position, where the force vanishes and, if it were initially at rest, it would remain at rest. Real particles obey the laws of quantum mechanics and this situation cannot occur.

The implications of a non-vanishing zero point energy are important in some instances. The atoms of a crystal, for example, oscillate with less energy as the temperature is lowered. Even at the lowest temperatures attainable, however, there is still some atomic motion.

The wave functions for the first few states, in order of increasing energy, are given in the following table. In these expressions $a=\left[h/(2\pi\sqrt{mk})\right]^{\tfrac{1}{2}}$, $b=a\sqrt{\pi}$, and the time dependent functions f_n all have magnitude 1. Their specific forms are not important for calculations of the probability density if the particle is in one of these states.

n	$\Psi(x,t)$
0	$(1/b)^{\tfrac{1}{2}} \, e^{-x^2/2a^2} \, f_0(t)$
1	$(1/2b)^{\tfrac{1}{2}} \, 2(x/a) \, e^{-x^2/2a^2} \, f_1(t)$
2	$(1/8b)^{\tfrac{1}{2}} \left[2 - 4(x/a)^2\right] e^{-x^2/2a^2} \, f_2(t)$

$$3 \qquad (1/48b)^{\frac{1}{2}} \left[12(x/a) - 8(x/a)^3 \right] e^{-x^2/2a^2} f_3(t)$$

$$4 \qquad (1/384b)^{\frac{1}{2}} \left[12 - 48(x/a)^2 + 16(x/a)^4 \right] e^{-x^2/2a^2} f_4(t)$$

Problem 6.   An electron has a potential energy function given by $V=61x^2$ J, with x in meters.

a.   The electron is in the harmonic oscillator state with n=0. Use the program of Fig. 2-4 to plot the probability density as a function of x. Plot points at 2×10^{-11} m intervals from $x = -2.4 \times 10^{-10}$ m to $x = +2.4 \times 10^{-10}$ m.

b.   Suppose now that the electron is in the state with n=2 and again plot the probability density. Plot points from $x = -3.2 \times 10^{-10}$ m to $x = +3.2 \times 10^{-10}$ m.

c.   If the electron obeyed the laws of classical mechanics, it would not go beyond the points where E=V. For each of the states above, calculate the value of x for which $E_n = V$ and mark these points on the appropriate graph.

The graphs show several noteworthy features. As the energy increases, the wave functions and the probability density show more spatial oscillations. In a very rough way, one can say that the particle has a larger momentum and the wave function has a shorter wavelength. This is not precise because the particle is not free and the wave function is not sinusoidal.

As the energy increases, the probability density extends to larger values of x. There is a greater probability of finding the particle at large x when the energy is large than when it is small. This phenomenon has an analogy in the motion of an oscillator which obeys classical laws. In that case, a higher energy means a greater amplitude and the excursion of the particle from equilibrium is greater.

The graphs also show that the particle has a chance of being found at places beyond the points where, classically, it would stop and then start back toward the equilibrium point. It has a chance of being found in regions where the potential energy is greater than the kinetic energy. In these regions, the wave function is not oscillatory but, rather, it decreases rapidly toward zero.

Problem 7.   An electron has a potential energy function given by $V(x)=61x^2$ J, with x in meters. It is in the harmonic oscillator state with n=4.

a. Find the values of x for which the probability density vanishes. Use a root find-ing program to search for the zeros of Ψ, not $|\Psi|^2$.

b. Find the values of x for which the probability density is a maximum. Differentiate the wave function with respect to x, then use a root finding program to find the zeros of the derivative.

c. Find the points where E=V(x). These are the classical turning points where the particle would stop and turn back if it obeyed the laws of classical mechanics.

d. Use the coordinates found in parts a, b, and c, along with the discussion of the previous paragraphs, to sketch the probability density as a function of x.

26.2   Average Values and Uncertainties

Suppose a particle has some specified wave function and that its position is measured. The experiment is repeated many times and a large collection of data, con-sisting of values of the x coordinate, is obtained.

Experimental data of this nature are sometimes characterized by two parameters. the average value and the standard deviation or uncertainty.

The average value is found by summing the individual values and dividing the sum by the number of data points. If N experiments are performed and x_i is the coordinate obtained in experiment i, then the average x_{ave} is given by

$$x_{ave} = (1/N)\sum_{i=1}^{N} x_i .$$
(26-9)

The uncertainty measures the spread in the data. It averages the square of the deviations of the various values from the average. It is denoted by Δx and its square is given by

$$(\Delta x)^2 = (1/N)\sum_{i=1}^{N}(x_i-x_{ave})^2 .$$
(26-10)

These ideas can be illustrated by means of a calculator simulated experiment, similar to the one discussed in connection with photon waves.

Problem 1.   An electron is confined to a box with sides at x=0 and x=1.6×10^{-8} m. Its position is measured when it is in the state with n=2. Eight particle detectors, each 2×10^{-9} m wide, are placed along the x axis. To simulate the measurement, select a number at random from those between 0 and 1000, inclusive, then use the following table to find which detector receives the electron.

Random Number	Detector Center
0 – 45	1×10^{-9} m
46 – 250	3×10^{-9} m
251 – 455	5×10^{-9} m
456 – 500	7×10^{-9} m
501 – 545	9×10^{-9} m
546 – 750	11×10^{-9} m
751 – 955	13×10^{-9} m
956 – 1000	15×10^{-9} m

a.   Obtain the value of the coordinate as measured in each of 150 experiments. Use the coordinate for the center of the detector which receives the particle. Draw a histogram showing the number of times a particle is found in each detector.

b.   Calculate the average x coordinate.

c.   Calculate the uncertainty in the x coordinate.

     The simulated experiment emphasizes the probabalistic nature of phenomena at the atomic level. Different experiments yield different results because the value of x obtained in an experiment is a matter of chance, not because the particle has moved since the last measurement. The experiment is restarted each time, with the particle in the original state.

     The average value of x and the uncertainty in x, in the limit of a large number of experiments, can be calculated directly from the probability distribution. Since $|\Psi|^2$ dx is the probability of finding the particle between x and x+dx, the average value of x is given by

$$x_{ave} = \int_{-\infty}^{+\infty} x \, |\Psi|^2 \, dx \qquad (26-11)$$

and the square of the uncertainty is given by

$$(\Delta x)^2 = \int_{-\infty}^{+\infty} (x - x_{ave})^2 \, |\Psi|^2 \, dx. \qquad (26\text{-}12)$$

For a particle in a box, $\Psi = 0$ outside the box and these expressions reduce to

$$x_{ave} = \int_0^L x |\Psi|^2 \, dx \qquad (26\text{-}13)$$

and

$$(\Delta x)^2 = \int_0^L (x - x_{ave})^2 \, |\Psi|^2 \, dx, \qquad (26\text{-}14)$$

respectively.

Problem 2.   An electron is confined to a box with sides at $x=0$ and $x=1.6\times10^{-8}$ m.   Its position is measured when it is the state with $n=2$.

a.   Use the program of Fig. 8-1, with $N=20$, to evaluate Eqn. 26-13 for x_{ave}.

b.   Use the program of Fig. 8-1, with $N=20$, to evaluate Eqn. 26-14 for $(\Delta x)^2$, then take the square root to find the uncertainty in x.

c.   Compare the answers found in parts a and b with the corresponding results of problem 1.   Agreement is not exact for two reasons.   First, the number of simulated experiments performed in working problem 1 is too small to qualify as approximating the limit required for Eqns. 26-13 and 26-14 to predict experimental results. Second, the detectors described in problem 1 are too wide for the sums performed there to approximate integrals.   As the number of experiments and the number of detectors are increased, the simulated experimental results approach those found in parts b and c.

Problem 3.   An electron has a potential energy function given by $V(x)=61x^2$ J, with x in meters.

a.   Look at the probability density functions for $n=0$ and $n=2$.   These were plotted in answer to problem 6 of section 26.1   For each of these states, what is the average of a large number of position measurements?   For which state is the uncertainty in position greater?   Answer these questions by noting the shapes of the curves and how far they extend along the x axis.

b.   For each of the states, use the program of Fig. 8-1 to calculate x_{ave} and Δx. Replace the infinite limits of the integrals with -5×10^{-10} m and $+5\times10^{-10}$ m, respectively and use $N=50$.

526

26.3 Superpositions of States

In the last two sections, some possible wave functions for a particle were studied and these were used to find the probability of finding the particle in specified regions of space. It is possible to measure other attributes of the particle, such as its energy and momentum. In this section, some aspects of the relationship between the wave function and these quantities are investigated.

In order to describe how the wave function can be used to predict the results of energy and momentum measurements, the mathematics of complex functions must be used. A complex function is composed of two parts, one of which is called the real part, the other of which is called the imaginary part.

The unit imaginary number is the square root of -1 and is denoted by i. It should not be confused with the unit vector in the x direction, which is denoted by $\hat{i}$. The number 5i, for example, is said to be imaginary and its square is -25. The number 7+3i is said to be complex. It consists of a real part (7) and an imaginary part (3). Its square is 49+42i-9=40+42i, another complex number.

If $f(x)$ is an ordinary real function of the coordinate x, then $if(x)$ is imaginary. The coordinate is real. If $f(x)$ is complex, it can be written $f(x)=f_R(x)+if_I(x)$, where f_R and f_I are real functions. f_R is called the real part of f and f_I is called the imaginary part.

Complex numbers and functions can be added. In general, the sum is complex, with the real part equal to the sum of the real parts of the addends and the imaginary part equal to the sum of the imaginary parts of the addends. Symbolically,

$$f_1 + f_2 = (f_{1R}+f_{2R}) + i(f_{1I}+f_{2I}). \tag{26-15}$$

Subtraction is defined in a similar manner, with minus signs replacing the plus signs in the parentheses.

A function, called the complex conjugate of $f(x)$, is found by replacing i with -i wherever it appears. It is $f_R(x)-if_I(x)$. The square of the magnitude of the function is the product of the function and its complex conjugate:

$$|f(x)|^2 = \left[f_R(x)+if_I(x)\right]\left[f_R(x)-if_I(x)\right]$$

$$= f_R^2(x) - i^2 f_I^2(x) = f_R^2(x) + f_I^2(x) \qquad (26\text{-}16)$$

where $i^2=-1$ was used. The magnitude squared is real.

In general, particle wave functions are complex functions and, when the probability is computed, Eqn. 26-16 must be used to find the square of the magnitude.

The complex function we shall use most frequently has the form

$$F(\alpha) = e^{i\alpha}. \qquad (26\text{-}17)$$

This function is mathematically identical to $F=\cos\alpha + i\sin\alpha$ and its magnitude squared is

$$|F|^2 = e^{i\alpha}\, e^{-i\alpha} = e^0 = 1. \qquad (26\text{-}18)$$

This may also be evaluated using $|F|^2=(\cos\alpha + i\sin\alpha)(\cos\alpha - i\sin\alpha)=\cos^2\alpha + \sin^2\alpha=1$. Other useful relationships are

$$\sin\alpha = (1/2i)(e^{i\alpha} - e^{-i\alpha}) \qquad (26\text{-}19)$$

and

$$\cos\alpha = \tfrac{1}{2}(e^{i\alpha} + e^{-i\alpha}). \qquad (26\text{-}20)$$

These follow when $e^{i\alpha}=\cos\alpha +i\sin\alpha$ and $e^{-i\alpha}=\cos\alpha -i\sin\alpha$ are subtracted and added respectively.

In section 26.1, the time dependent part of a wave function was included by multiplying the coordinate dependent part by an unspecified function $f_n(t)$. We can now specify the function. It is

$$f_n(t) = e^{-i\omega t} \qquad (26\text{-}21)$$

where $\omega=2\pi E_n/h$ and E_n is the energy of the particle when it is in the state n. Since $e^{-i\omega t}=\cos(\omega t)-i\sin(\omega t)$, both the real and imaginary parts of $e^{-i\omega t}$ oscillate with angular frequency ω. For particles, the frequency of oscillation of the wave function is related to the energy of the particle in the same way as for photons.

528

One reason why complex functions must be used should now be clear. When the particle is in a state with energy E, its wave function oscillates in time with angular frequency $\omega = 2\pi E/h$, but the probability density is independent of the time. The function $e^{-i\omega t}$ satisfies these requirements, while the functions $\sin(\omega t)$ and $\cos(\omega t)$ do not.

The wave functions studied in the last two sections oscillate sinusoidally with a frequency determined by the particle energy. Such wave functions are termed energy wave functions for the particle and the states are called energy states.*

A wave function may be a combination of energy wave functions. For a particle in a box, a possible wave function is given by

$$\Psi(x,t) = A(2/L)^{\frac{1}{2}}\sin(r\pi x/L)e^{-i2\pi E_r t/h}$$

$$+ B(2/L)^{\frac{1}{2}}\sin(s\pi x/L)e^{-i2\pi E_s t/h}, \qquad (26\text{-}22)$$

a combination of the n=r and n=s energy wave functions. The coefficients A and B are constants and satisfy the relationship $|A|^2 + |B|^2 = 1$.

The probability density is now time dependent:

$$|\Psi|^2 = A^2(2/L)\sin^2(r\pi x/L) + B^2(2/L)\sin^2(s\pi x/L)$$

$$+ (4/L)AB\sin(r\pi x/L)\sin(s\pi x/L)\cos\left[2\pi(E_r-E_s)t/h\right] \qquad (26\text{-}23)$$

where, for simplicity, we have taken A and B to be real, although they need not be, and have used $e^{i\alpha}+e^{-i\alpha}=2\cos\alpha$. Notice that the probability density oscillates with frequency $(E_r-E_s)/h$, a quantity which is proportional to the difference in the energies of the two states.

When it is in this state, the particle seems to have two energies, E_r and E_s, and some clarification of the meaning of the wave function is required. The interpretation is again in terms of probability. When the energy of the particle is measured, the result is either E_r or E_s, not some combination of the two. The probability that E_r is the measured value is $|A|^2$, the magnitude squared of the coefficient multiplying the

---

* Strictly speaking, these functions are called energy eigenfunctions and the states are called energy eigenstates. The word eigen means characteristic in German.

energy function associated with E_r. Similarly, the probability that E_s is the measured value is $|B|^2$, the magnitude squared of the coefficient multiplying the energy function associated with E_s.

The situation being discussed should be contrasted to those described in the last two sections. There, the particle wave functions were assumed to oscillate sinusoidally with time and a single frequency was associated with each of them. Such a particle has a single, definite energy associated with it and every energy measurement, taken with the particle in the same state, produces the same result. The probability density is independent of time.

Problem 1.   An electron is confined to the segment of the x axis between x=0 and $x=1.6\times 10^{-8}$ m. Its wave function is a combination of the energy wave functions with n=1 and n=2. Take the coefficient for the n=1 state to be 0.6 and the coefficient for the n=2 state to be −0.8 in Eqn. 26-22.

a.   Calculate the two possible results of an energy measurement and find the probability for each to occur.

b.   For each of the following values of the time, use the program of Fig. 2-4 to plot the probability density as a function of x. Use 1×10^{-9} m intervals.

i.   t=0.

ii. $t=\frac{1}{4}h/(E_2-E_1)$.

iii. $t=\frac{1}{2}h/(E_2-E_1)$.

c.   Calculate the shortest time it takes for the probability density to return to the form it has at t=0.

d.   For each of the times listed in part b, use the program of Fig. 8-1, with N=20, to find the probability that the particle is found in the region from $x=3\times 10^{-9}$ m to $x=5\times 10^{-9}$ m.

By looking at the graphs, you should be able to visualize the probability density as it changes with time. The function consists of two peaks, one on the right side of the box and the other on the left side. The peaks do not move across the box. They remain at the same places, but change in size. At first, the peak on the right is large while the peak on the left is small. There is a high probability of finding the particle on the right side of the box.

As time goes on, the right peak decreases and the left one increases in size.

530

At the second value of the time considered, there is an equal probability of finding the particle on either side of the box center. At the last value of the time considered, there is a high probability of finding it on the left side.

For each of the individual energy wave functions, the probability of finding the particle in the region from $x=3\times10^{-9}$ m to $x=5\times10^{-9}$ m was calculated in problem 4 of section 26.1. Notice that the probability for the combination is not the same as the probability for either constituent state nor is it the sum of the probabilities.

The probability of finding the particle in the selected segment of the x axis changes with time. Again the graphs of part a are revealing. At t=0, the peak on the left side of the box is small while that on the right is large. Since the selected segment is on the left side, we expect a small probability of finding the particle there. Later, when the left peak is larger, so is the probability of finding the particle in the selected segment.

The average value and the uncertainty for a series of coordinate measurements are also time dependent when the wave function has a form like that given by Eqn. 26-22.

Problem 2. For the situation described in problem 1, and for each of the values of the time given there, calculate the average value of x and the uncertainty Δx. Use the program of Fig. 8-1, with N-30, to evaluate the integrals which appear in Eqns. 26-13 and 26-14. After completing the calculations for the first two values of the time, guess x_{ave} and Δx for the last value. Then perform the calculations for that value. On each of the graphs of problem 1, mark the points x_{ave} and $x_{ave}\pm\Delta x$.

Care must be taken in the experimental interpretation of the average value and the uncertainty. Each coordinate measurement is taken at the time t, measured from t=0, when the particle is placed in the state. In each case, the wave function at the start of the experiment must be given by Eqn. 26-22 with t=0. The average value, for example, does not change with time because the particle moves between successive measurements. The calculation has nothing to say about the particle after a coordinate measurement is taken.

In practice, the coefficients A and B in Eqn. 26-22 are usually determined by the initial conditions, the wave function at t=0. These, in turn, are determined by the experimental procedure: how the particle is placed in the box, for example.

To give an example for which only two energy states are involved is difficult, but examples for which many such states must be taken into account are numerous. Suppose the particle is initially placed in the box so that it has a uniform probability of being anywhere within a small region of the x axis near x=0. Its initial wave function is then a constant in this small region and zero elsewhere. This wave function can be constructed as the superposition of a large number of energy wave functions. The coefficients of these functions are numbers chosen so that, when the energy wave functions are summed for t=0, the form of the initial wave results. Then, as time goes on, each energy wave function oscillates with a different frequency and the total probability density changes.

We can now say a little more about the measurement process. If the particle has the wave function given by Eqn. 26-22 and its energy is measured, either E_r or E_s is the result. Which value is actually obtained is a matter of chance and the probabilities are given by the squares of the coefficients in the wave function. Now suppose, in one experiment, the energy is found to be E_r. Once the measurement is taken, the wave function for the particle is no longer given by Eqn. 26-22 but, instead, it is the energy wave function corresponding to energy E_r: $(2/L)^{\frac{1}{2}}\sin(r\pi x/L)\exp(-i2\pi E_r t/h)$. The act of measuring the energy changes the wave function and now we see what the change is.

If, after the first energy measurement, a second one is performed without resetting the wave function, the result is exactly the same as the first. For the example given above, if E_r is obtained for the first measurement, then E_r is also obtained for the second. After the first measurement, the particle wave function no longer contains the E_s component and the probability of obtaining E_s is 0.

We now turn to the study of momentum. The function

$$\Psi = A\, e^{i(kx-\omega t)}, \qquad (26\text{-}24)$$

where A, k, and ω are constants, represents a plane wave traveling in the positive x direction. It has amplitude A, angular frequency ω, and propagation constant $k=2\pi/\lambda$. Both the real and imaginary parts of Ψ are sinusoidal waves. The amplitude is determined by the initial conditions.

Such a wave is associated with a particle when the particle has momentum $p=hk/2\pi$

532

and energy $E = \hbar\omega/2\pi$. For this wave

$$|\Psi|^2 = |A|^2 \, e^{i(kx-\omega t)} \, e^{-i(kx-\omega t)} = |A|^2, \tag{26-25}$$

a constant. If the x axis is divided into segments of equal length, the probability of finding the particle in any particular segment is the same, regardless of which segment is chosen.

This result is found experimentally. If the particle has momentum p and energy E, its wave must be a sinusoidal wave with angular frequency $\omega = 2\pi E/h$ and wavelength $\lambda = p/h$. But the wave must also predict equal probabilities for equal segments. It is clear that the real wave function $A \cos(kx-\omega t)$ does not do the job but the complex wave function does.

Wave functions can be constructed by summing various momentum functions. For example

$$\Psi(x,t) = A \, e^{i(2\pi p_1 x/h \, - \, \omega_1 t)} + B \, e^{i(2\pi p_2 x/h \, - \, \omega_2 t)}, \tag{26-26}$$

where A and B are constants, is a possible wave function for a particle. The relationship $k = 2\pi p/h$ was used in writing this wave function.

If the momentum of the particle is measured, the result is either p_1 or p_2. The probability of obtaining p_1 is proportional to $|A|^2$, while the probability of obtaining p_2 is proportional to $|B|^2$.

If the potential energy function vanishes everywhere, then the energy of the particle is wholly kinetic energy and $E_1 = p_1^2/2m$ where m is the mass of the particle. Since $\omega_1 = 2\pi E_1/h$, it follows that $\omega_1 = \pi p_1^2/mh$. A similar expression holds for ω_2. If the potential energy function does not vanish everywhere, a different analysis must be used.

For most wave functions which are used in practical problems, the possible values of the momentum are not discrete, as they were for the example given above. Instead, they form a continuum and the sum of the momentum wave functions is, in reality, an integral. The wave function is then written

$$\Psi(x,t) = h^{-\frac{1}{2}} \int_{-\infty}^{+\infty} f(p) \, e^{i(2\pi px/h \, -\omega t)} \, dp. \tag{26-27}$$

The quantity $\left|f(p)\right|^2 dp$ is the probability that a momentum measurement yields a value between p and p+dp. This is the magnitude squared of the coefficient of $h^{-\frac{1}{2}}e^{i(2\pi px/h - \omega t)}$ in the "sum" which represents the wave function. Since every measurement yields some value of p, the probabilities must add to 1, and f(p) must satisfy

$$\int_{-\infty}^{+\infty} \left|f(p)\right|^2 dp = 1. \qquad (26\text{-}28)$$

The factor $h^{-\frac{1}{2}}$ appears in Eqn. 26-27 so that $\int_{-\infty}^{+\infty}\left|\Psi\right|^2 dx=1$ if f(p) satisfies Eqn. 26-28.

If a large number of momentum measuring experiments are conducted, the quantity $\left|f(p)\right|^2 dp$ gives the fraction for which the result is between p and p+dp. Each experiment must start with the particle in the same state. This is an important consideration since a momentum measurement, in general, changes the state of the particle and with it the particle wave function. In fact, after the measurement, the wave function has the form given in Eqn. 26-24.

In order to demonstrate some properties of wave functions, we consider a simple case. At t=0, the wave function we consider is given by

$$\Psi(x,0) = A\int_{-\infty}^{+\infty} e^{-\alpha p^2}\, e^{i2\pi px/h}\, dp, \qquad (26\text{-}29)$$

where A and α are constants. In the next two problems, the momentum distribution is examined. After that, the position probability distribution is investigated.

Problem 3. At t=0, a particle wave function is given by

$$\Psi(x,0) = (1/h\delta\sqrt{\pi})^{\frac{1}{2}}\int_{-\infty}^{+\infty} e^{-0.5(p/\delta)^2}\, e^{i2\pi px/h}\, dp$$

where $\delta=5.2\times10^{-25}$ kg·m/s.

a. Use the program of Fig. 2-4 to plot the momentum probability density $(1/\delta\sqrt{\pi})e^{-(p/\delta)^2}$ as a function of p. Plot points every 1×10^{-25} kg·m/s from -12×10^{-25} kg·m/s to $+12\times10^{-25}$ kg·m/s. The function is symmetric. Calculate values for p$\geq$0 only.

b. On the graph, mark the maximum of the distribution. Values of the momentum in the neighborhood of this point occur most often when the momentum is repeatedly measured, in different experiments.

534

c. Use the graph to find the probability per unit momentum at the points where $p=-\delta$ and $p=+\delta$. This should be $e^{-1}\approx0.37$ times the value at the maximum. The parameter δ is a measure of the width of the momentum probability distribution. If δ is large there is likely to be a large spread in the momentum values obtained in a series of experiments. If δ is small the spread is likely to be small.

d. Show that the momentum probabilities sum to 1. Use the program of Fig. 8-1 to evaluate Eqn. 26-28. The machine cannot handle the infinite integration limits. Instead of those, use 0 and $+5\times10^{-24}$ kg·m/s as the lower and upper limits, respectively, then double the result. At the upper limit, the integrand is extremely small and the error introduced by this approximation is negligible. Use N=20.

For momentum measurements, the function $f(p)$ plays the same role as the wave function $\Psi(x,t)$ does for position measurements. The probability that a momentum measurement gives a value between p_1 and p_2 is given by the integral

$$P = \int_{p_1}^{p_2} |f(p)|^2 \, dp. \tag{26-30}$$

If momentum measuring experiments are performed, always starting with the particle in the same state, the result is a collection of momentum values. It is sometimes of interest to know the average value and the standard deviation or uncertainty of this data. In the limit of a large number of experiments, the average is given by

$$p_{ave} = \int_{-\infty}^{+\infty} p \, |f(p)|^2 \, dp \tag{26-31}$$

and the square of the uncertainty is given by

$$(\Delta p)^2 = \int_{-\infty}^{+\infty} (p-p_{ave})^2 \, |f(p)|^2 \, dp. \tag{26-32}$$

These equations should be compared with Eqns. 26-11 and 26-12, respectively.

Problem 4. A particle has the same wave function as the particle of problem 3. Use the program of Fig. 8-1, with N=20, to calculate the following quantities. When appropriate, replace the infinite integration limits by -5×10^{-24} kg·m/s and $+5\times10^{-24}$ kg·m/s respectively. The momentum probability density is

$$|f(p)|^2 = (1/\delta\sqrt{\pi})e^{-(p/\delta)^2}.$$

a. The probability that a momentum measurement produces a value between $p=2\times10^{-25}$ kg·m/s and $p=4\times10^{-25}$ kg·m/s.

b. The average value of the results of a large number of momentum measurements, all taken with the particle in the state specified.

c. The uncertainty in the momentum, in the limit of a large number of experiments. Take the integration limits to be 0 and $+5\times10^{-24}$ kg·m/s, then double the result. Compare Δp with the value of δ. For functions of the form given, the exact relationship is $\Delta p=\delta/\sqrt{2}$.

We now go on to investigate the probability density for the particle coordinate x. The function $\Psi(x,0)$, given by Eqn. 26-29, can be evaluated more easily if it is put in a slightly different form. The integration interval is broken into two parts, one running from $-\infty$ to 0 and the other running from 0 to $+\infty$. In the first integral, p is replaced by $-p$, with the appropriate change in the sign of the lower limit. The limits on the first integral are interchanged and the sign in front changed. The two integrals can now be combined. The sequence of steps is

$$\int_{-\infty}^{+\infty} e^{-\alpha p^2} e^{i2\pi px/h}\,dp = \int_{-\infty}^{0} e^{-\alpha p^2} e^{i2\pi px/h}\,dp + \int_{0}^{+\infty} e^{-\alpha p^2} e^{i2\pi px/h}\,dp$$

$$= -\int_{+\infty}^{0} e^{-\alpha p^2} e^{-i2\pi px/h}\,dp + \int_{0}^{+\infty} e^{-\alpha p^2} e^{i2\pi px/h}\,dp$$

$$= \int_{0}^{+\infty} e^{-\alpha p^2} e^{-i2\pi px/h}\,dp + \int_{0}^{+\infty} e^{-\alpha p^2} e^{i2\pi px/h}\,dp$$

$$= \int_{0}^{+\infty} e^{-\alpha p^2} \left[e^{i2\pi px/h} + e^{-i2\pi px/h}\right]\,dp$$

$$= 2\int_{0}^{+\infty} e^{-\alpha p^2} \cos(2\pi px/h)\,dp. \qquad (26\text{-}33)$$

536

In the last step, use was made of the identity $2\cos\theta = e^{i\theta} + e^{-i\theta}$. Finally

$$\Psi(x,0) = 2A \int_0^\infty e^{-\alpha p^2} \cos(2\pi px/h)\, dp. \qquad (26\text{-}34)$$

Problem 5.   A particle is in a state with wave function given by

$$\Psi(x,0) = (4/h\delta\sqrt{\pi})^{\frac{1}{2}} \int_0^\infty e^{-0.5(p/\delta)^2} \cos(2\pi px/h)\, dp,$$

with $\delta = 5.2\times10^{-25}$ kg·m/s. This is the same as the wave function for problem 3.

a.   Use the program of Fig. 8-1, with N=30, to evaluate the integral for various values of x, then plot $|\Psi|^2$ as a function of x. Plot points every 5×10^{-11} m from $x=-4\times10^{-10}$ m to $x=+4\times10^{-10}$ m. Since $\Psi(-x,0)=\Psi(x,0)$, you need perform the calculation only for $x\geq0$. Plot the function for both positive and negative values of x, however. Replace the upper limit with 5×10^{-24} kg·m/s.

b.   The probability density is a function of the form $A\exp(-\alpha x^2)$. Use the graph to estimate the values of x for which the ratio of $|\Psi(x,0)|^2$ to its value at x=0 is $e^{-1}\approx0.37$. What is the uncertainty Δx in the particle position? See part c of the last problem for the relationship between Δx and α.

For a particle with the given wave function, both position and momentum measurements yield a range of values. This statement is true for nearly all situations which occur, not just those for which the wave function is like that considered in the preceding problem.

It is possible to place the particle in a state for which the distribution of measured momentum values is narrower than that considered above. If this is done, however, the range of measured coordinate values broadens.

This is also true for other situations. In fact, it is a general principle that the product of the uncertainty in x and the uncertainty in p can never be smaller than $h/4\pi$. This is called the Heisenberg uncertainty principle and, in mathematical terms, it is written

$$(\Delta x)(\Delta p) \geq h/4\pi. \qquad (26\text{-}35)$$

<u>Problem 6.</u>   A particle wave function has the same form as that given in problem 3 but with $\delta = 2.6 \times 10^{-25}$ kg·m/s.

a.   Find the uncertainty Δp in the momentum. Use the program of Fig. 8-1, with N=30, to evaluate the integral which appears in Eqn. 26-32. For this wave function $p_{ave} = 0$. Since the integrand is an even function, you may take the lower limit to be 0, then double the result. Take the upper limit to be 3×10^{-24} kg·m/s.

b.   Plot $|\Psi(x,0)|^2$ as a function of x from $x = -8 \times 10^{-10}$ m to $x = +8 \times 10^{-10}$ m. Use 1×10^{-10} m intervals.

c.   Use the graph to estimate the value of x for which the ratio of $|\Psi(x,0)|^2$ to its value at x=0 is $e^{-1} \simeq 0.37$. Calculate Δx.

d.   Evaluate the product $(\Delta x)(\Delta p)$ and compare the result with $h/4\pi$. Do the same for Δx and Δp as found in problems 4 and 5.

The ideas which have been presented here are quite general in their applicability. Any wave function can be represented as the sum or integral of plane waves, each of the form given in Eqn. 26-24. General mathematical techniques exist for finding f(p), given $\Psi(x,t)$.

It is now easy to see how the spatial wave function $\Psi(x)$ is used to find the probability distribution for momentum. First the function f(p) which satisfies Eqn. 26-27 is found, then $|f(p)|^2$ is evaluated.

You should also be aware of the similarity in the techniques presented to calculate the probability that a specified value of the energy is measured and to calculate the probability that specified values of momentum are measured. In each case, the wave function is first written as a linear superposition of functions which are characteristic of the quantity to be measured: energy wave functions or momentum wave functions. Then the coefficients of the functions are used to calculate the probabilities.

The wave function can be thought of as a sum of probability amplitude waves for energy or as a sum of probability amplitude waves for momentum. Which sum is used depends on the quantity to be measured.

There are many other quantities which are of interest and they all have characteristic wave functions. Any wave function can be written as a linear superposition of the characteristic waves for a particular quantity and the coefficients squared to give the probabilities for obtaining various values of that quantity, when it is measured.

Chapter 27

THE SCHRÖDINGER EQUATION

A numerical technique is presented for solving the time independent Schrödinger equation for characteristic energy wave functions. Solutions for several low lying states of the particle in a box and the harmonic oscillator are obtained and compared with the functions given in the last chapter. In the second section, various one dimensional problems are solved and the solutions are used to draw qualitative conclusions about the influence of the potential energy function on the wave function and energy levels. This material extends the coverage of quantum mechanics found in Chapter 50 of PHYSICS, Chapter 43 of FUNDAMENTALS OF PHYSICS, and Chapter 6 of RELATIVITY AND EARLY QUANTUM THEORY. Three of the problems, on barrier tunneling, can be solved in conjunction with Chapter 47 of FUNDAMENTALS OF PHYSICS (EXTENDED).

27.1   Integration of the Schrödinger Equation

One dimensional wave functions for non-relativistic particles with a definite energy obey the differential equation

$$-(h^2/8\pi^2 m) \ d^2\psi/dx^2 + V(x)\psi = E\psi, \qquad (27-1)$$

where m is the mass of the particle, h is Planck's constant, and E is the energy of the particle. $V(x)$ is the potential energy function, a function of x.

The solution to this equation is independent of time and is denoted by $\psi(x)$. It can be multiplied by $e^{-i\omega t}$ to obtain the corresponding time dependent function $\Psi(x,t)$. Here $\omega = 2\pi E/h$.

Eqn. 27-1, called the time independent Schrödinger equation, is a fundamental equation for non-relativistic quantum mechanics. Using this equation, it is possible to solve for the particle wave function, given its potential energy function. We shall use it to explore the influence of the potential energy function on the wave function.

For a given potential energy function, there exists a family of solutions to the differential equation. The correct solution for a given physical situation is selected by application of boundary conditions. To do this, some information must be known about

538

the wave function for some values of x.

If there is an impenetrable rigid wall at some value of x, the wave function vanishes there. This is an example of a boundary condition and it is written $\psi(x_0)=0$, where x_0 is the coordinate of the wall.

Most problems do not involve rigid walls and the wave function can extend to all values of x. For bound particles, the usual boundary conditions are given in the form of the limits $\psi \to 0$ as $x \to \infty$ and $\psi \to 0$ as $x \to -\infty$. These conditions eliminate solutions for which the particle has high probability of being far from the origin. In most cases, they also eliminate solutions for which $\int |\psi|^2 \, dx$ blows up. Since the integral gives a probability, it must be finite no matter what the limits of integration are.

The boundary conditions are extremely important for the solution of problems. Not only are they needed in order to obtain the correct wave function, but, in the case of bound particles, they are needed to obtain the correct values of the allowed energy levels.

This last remark merits some discussion. If the particle is bound, the Schrödinger equation has solutions which meet the boundary conditions only if the energy E is one of a set of discrete values. That is, the energy is quantized and E must have one of the allowed values. If it does not, the only solutions are those which do not meet the boundary conditions or else $\psi=0$ everywhere.

In the Schrödinger equation, the energy E of a bound particle is unknown. For a given potential energy function, not only is the wave function sought but so is the energy. The object of most problems is to find a value for E such that a solution ψ meets the boundary conditions, and to integrate the Schrodinger equation to obtain the function ψ.

This can be done numerically. Schrödinger's equation is a second order differential equation, just as is the equation for Newton's second law. The independent variable is now x, rather than the time and the dependent variable is now the wave function ψ, rather than the particle coordinate. The value of ψ and its derivative $d\psi/dx$ must be given for some initial value of x, then the program of Fig. 6-3 can be used to produce ψ at a series of points along the x axis.

Rather than use the program of Fig. 6-3, however, we present a similar program, written in terms appropriate for the Schrödinger equation. Additionally, the probability density $|\psi|^2$ is computed and displayed, while $d\psi/dx$ is not displayed. These revisions are incorporated into the program.

For purposes of computation, the Schrödinger equation is rearranged a bit and written

$$d^2\psi/dx^2 = f(x) \tag{27-2}$$

where $f(x)=(8\pi^2 m/h^2)(V-E)\psi$. The technique used to solve this equation is similar to that used to solve the Newton's second law equation. The x axis is divided into intervals of uniform width Δx, with interval j running from x_{j-1} to x_j. The intervals must be so narrow that f(x) can be taken to be constant in any one of them, although different values are used for different intervals. By integrating Eqn. 27-2 twice, the value of ψ at the end of interval j can be found in terms of the values of ψ and its first derivative at the beginning of the interval. These last two quantities are the same as ψ and $d\psi/dx$, respectively, evaluated at the end of interval j-1 and, consequently, they are labelled ψ_{j-1} and $(d\psi/dx)_{j-1}$ respectively.

Let ψ_j be the value of the wave function at the end of interval j and $(d\psi/dx)_j$ be the value of the derivative at the end of the same interval. Then integration of Eqn. 27-2 yields

$$\psi_j = \psi_{j-1} + (d\psi/dx)_{j-1}\Delta x + \tfrac{1}{2}f(x_{j-1})(\Delta x)^2$$

$$= \psi_{j-1} + \left[(d\psi/dx)_{j-1} + \tfrac{1}{2}f(x_{j-1})\,\Delta x\right]\Delta x \tag{27-3}$$

where, in the last form, Δx has been factored from the last two terms. To obtain this equation, f is assigned the value it has at the beginning of the interval and it is assumed to be constant throughout the interval.

The quantity in the square brackets of Eqn. 27-3 is just the value of the derivative at the midpoint of interval j. The value of f multiplied by half the interval width is added to the value of the derivative at the beginning of the interval. If we write $(d\psi/dx)_{j\,\frac{1}{2}}$ for the derivative evaluated at the midpoint of interval j, then a single integration of Eqn. 27-2 from the beginning to the midpoint of interval j yields

$$(d\psi/dx)_{j\ \frac{1}{2}} = (d\psi/dx)_{j-1} + \tfrac{1}{2}f(x_{j-1})\ \Delta x, \tag{27-4}$$

the quantity in the square brackets of Eqn. 27-3. Substitution of this expression into Eqn. 27-3 gives

$$\psi_j = \psi_{j-1} + (d\psi/dx)_{j\ \frac{1}{2}}\ \Delta x. \tag{27-5}$$

It is convenient to calculate $d\psi/dx$ at the midpoint of an interval in terms of its value at the midpoint of the previous interval. The relationship is given by an equation which is similar to Eqn. 27-4 except that the factor $\frac{1}{2}$ does not appear since the two points are a full interval apart. The expression is

$$(d\psi/dx)_{j\ \frac{1}{2}} = (d\psi/dx)_{j-1\ \frac{1}{2}} + f(x_{j-1})\ \Delta x. \tag{27-6}$$

Here f is evaluated at the beginning of interval j and is assumed to have the same value throughout the range from the midpoint of interval j-1 to the midpoint of interval j.

Given ψ at the beginning of an interval and $d\psi/dx$ at the midpoint of the previous interval, Eqn. 27-6 is used to calculate $d\psi/dx$ at the midpoint, then Eqn. 27-5 is used to calculate ψ at the end of the interval. This process is repeated until the final value of x is reached.

To start the calculation, ψ and $d\psi/dx$ are entered for the initial value of x. The first job is to calculate $d\psi/dx$ for the point half an interval before the initial point. This is done to start the sequence of $d\psi/dx$ calculations. To do it,

$$(d\psi/dx)_{0\ \frac{1}{2}} = (d\psi/dx)_0 - \tfrac{1}{2}f(x_0)\ \Delta x \tag{27-7}$$

is used.

The flow chart is shown in Fig. 27-1. It uses the following storage locations:

X1: ψ,
X2: $d\psi/dx$,
X3: x,
X4: Δx,
X5: N,
X6: f,

and

X7: counter.

542

Figure 27-1. Flow chart for a program to integrate the Schrodinger equation.

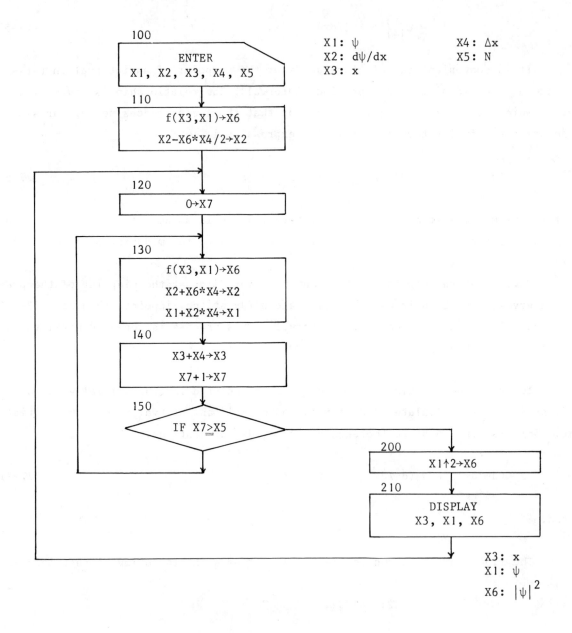

At line 100, ψ, $d\psi/dx$, and x for the initial point are entered. So is the interval width Δx and the number of intervals N to be considered before the machine displays results. At line 110, $d\psi/dx$ at $x=x_0-\frac{1}{2}\Delta x$ is computed and stored in X2. At line 130, a loop is entered. It is traversed N times and when finished, the value of ψ at the end of N intervals is displayed, at line 210. First, $|\psi|^2$ is computed, at line 200, and it is also displayed. Then the machine returns to line 120 to consider the next N intervals. X7 is a counter used to count the number of times the loop is traversed.

The heart of the loop is at line 130. There, $d\psi/dx$ at the midpoint of the interval and ψ at the end point are computed and stored in X2 and X1 respectively. Eqns. 27-5 and 27-6 are used.

At line 140, x is incremented by Δx so that it has the value of the coordinate for the end of the interval. This is the value for which ψ was computed at line 130. The counter is also incremented. If, at line 150, the counter has not yet reached N, the machine does not display results but returns immiediately to line 130 to consider the next interval.

It is important to take Δx small so that the error made in assuming f is constant in an interval does not produce a large error in ψ. For most of the problems Δx is chosen so that ψ is correct to two significant figures. This is accurate enough for plotting purposes, but you may desire more accuracy. If you do, reduce Δx by a factor of two or more. The running time is increased proportionately.

Before starting to calculate wave functions, there is a technicality which must be discussed. In general, the functions produced by the program do not satisfy $\int|\psi|^2\, dx = 1$, as they should to be true wave functions. However, once a solution has been obtained, it can be multiplied by any constant without losing its status as a solution to Eqn. 27-1. The constant is picked so that the integral of the probability density over the entire x axis is 1. The wave function is then said to be normalized.

This property of the solutions shows up in the way the differential equation is solved. Generally, either ψ or $d\psi/dx$ is known at some value of x, but not both. The other is chosen arbitrarily. Two people picking different starting values obtain different solutions, but they differ only in the multiplicative constant and they become the same once they are normalized. Except for the first few problems in this section, we shall be satisfied with unnormalized wave functions.

544

In order to test the program, the first problem deals with the computation of wave functions for the low lying energy states of a particle in a box. Assume the energy, given by Eqn. 26-3, is known. The differential equation can then be simplified somewhat. Substitution of Eqn. 26-3 and V=0 into Eqn. 27-1 yields

$$d^2\psi/dx^2 = -(n\pi/L)^2\psi \qquad (27-8)$$

and the function f which appears in the program is calculated using

$$-(n\pi/L)\uparrow 2*X1\rightarrow X6$$

Problem 1. An electron is confined to the region of the x axis between x=0 and x=1.6×10^{-8} m. It is in the state with n=1 and has energy given by Eqn. 26-3. Use the program of Fig. 27-1 to integrate Eqn. 27-8 for $\psi(x)$. Take $\psi=0$ and $d\psi/dx=2.2\times10^{12}$ m$^{-3/2}$ at x=0. Use an interval width of $\Delta x=1\times10^{-10}$ m and display ψ every 1×10^{-9} m. Stop when x=1.6×10^{-8} m.

a. Make tables of ψ and $|\psi|^2$ as functions of x.

b. Use the following approximate procedure to normalize the probability distribution. Sum all the values of $|\psi|^2$ in the table, then multiply the result by the interval width, 1×10^{-9} m. FInally, create a new table of the normalized probability density. Divide each of the old values of $|\psi|^2$ by the number you just found. The sum of the new values of $|\psi|^2$, when multiplied by 1×10^{-9}, produces 1.

c. As a check on accuracy, compute $(2/L)\sin^2(\pi x/L)$ at x=4×10^{-9} m, 8×10^{-9} m, 12×10^{-9} m, and 1.6×10^{-8} m. Compare the values with those found in part b.

In the above problem, the initial value of $d\psi/dx$ was chosen to be close to the correct value for the normalized wave function and the process of normalization did not change the values of $|\psi|^2$ much. In the next problem, we pick the initial value of $d\psi/dx$ to be greatly different from the correct value for the normalized wave function. Normalization then brings the values of $|\psi|^2$ into agreement with $(2/L)\sin^2(n\pi x/L)$.

Problem 2. An electron is confined to a box with sides at x=0 and x=1.6×10^{-8} m. It is in the state with n=2. The energy is given by Eqn. 26-3. Use the program of Fig. 27-1 to integrate Eqn. 27-8 and obtain $\psi(x)$ and $|\psi(x)|^2$. Take $\psi=0$ and $d\psi/dx=2.2\times10^{12}$ m$^{-3/2}$ at x=0. Use $\Delta x=1\times10^{-10}$ m and display ψ and $|\psi|^2$ every 1×10^{-9} m.

Stop when $x=1.6\times10^{-8}$ m.

a. Make a table of the values of ψ and $|\psi|^2$.

b. Normalize the probability distribution. Sum the values of $|\psi|^2$, then multiply the sum by 1×10^{-9} m. Divide each value of $|\psi|^2$ in the table by the result.

c. Check the accuracy of your results by comparing the normalized values of $|\psi|^2$ with $(2/L)\sin^2(2\pi x/L)$ at $x=8\times10^{-9}$ m, 1×10^{-8} m, 1.2×10^{-8} m, 1.4×10^{-8} m, and 1.6×10^{-8} m.

For the preceding two problems, the correct value for the energy was used. These values are known, of course, only because the problems have already been solved. We now assume the energy is not known and use the boundary conditions to solve for its value. The boundary conditions are $\psi=0$ at $x=0$ and $x=L$. The first is automatically satisfied when the initial value of ψ is entered. The energy E must be chosen so that the second condition is met.

The technique is simple. Guess a value for the energy, then integrate Eqn. 27-2, starting at $x=0$. When $x=L$ is reached, the value of ψ obtained will not, in general, be zero as it should be to meet the boundary condition there. Eqn. 27-2 is then integrated again, with a different value for the energy. The procedure is repeated until the value of the energy is found so that ψ is zero at $x=L$. In practice, the procedure is stopped when E has been found with acceptable accuracy.

The hunt for the value of the energy can be systematized somewhat. Suppose solutions to Eqn. 27-2 are found for a series of energy values, close together. We then have the values of ψ at $x=L$ for various values of E and $\psi(L)$ can be plotted as a function of E. Suppose, at small E, $\psi(L)$ is positive. Then as E increases, $\psi(L)$ decreases, passes through zero, and then increases negatively. The lowest allowed energy is the root of $\psi(L)$ as a function of E.

At higher values of E, $\psi(L)$ becomes less negative, passes through zero, and then becomes positive. As E increases, $\psi(L)$ oscillates from positive to negative and back again. It becomes zero at each allowed value of the energy.

To solve for the particle energy, $\psi(L)$ is fit to a polynomial in E, then the root of the polynomial is sought. For our purposes, linear interpolation is sufficiently accurate. If ψ_1 and ψ_2 are two values of the wave function at $x=L$, the first for energy E_1 and the second for energy E_2, then $\psi(L)$ for any value of E can be approximated by

546

$$\psi(L) = \psi_1 + \frac{\psi_2 - \psi_1}{E_2 - E_1}(E - E_1).\tag{27-9}$$

This is zero for

$$E = E_1 - \frac{E_2 - E_1}{\psi_2 - \psi_1}\psi_1.\tag{27-10}$$

To obtain higher accuracy, $\psi(L)$ for this value of E is found by integrating Eqn. 27-2, then this value replaces either ψ_1 or ψ_2 in Eqn. 27-10 and the allowed energy is recalculated. For best results, ψ_1 and ψ_2 should have opposite signs so the root lies between E_1 and E_2.

Problem 3. An electron is confined to a box with sides at x=0 and x=1.6×10^{-8} m. For the following integrations, use $\Delta x=5\times10^{-11}$ m and take $\psi=0$, $d\psi/dx=2\times10^{12}$ m$^{-3/2}$ at x=0.
a. Integrate Eqn. 27-2 to obtain $\psi(L)$ for E=2.1×10^{-22} J. Call this ψ_1.
b. Integrate Eqn. 27-2 to obtain $\psi(L)$ for E=2.5×10^{-22} J. Call this ψ_2.
c. Use Eqn. 27-10 to obtain an approximate value for the allowed energy level.
d. Integrate Eqn. 27-2 to obtain $\psi(L)$ for the value of E found in part c. Since $\psi(L)$ is negative, call it ψ_2. ψ_1 is still the value found in part a.
e. Use Eqn. 27-10 to obtain a better approximation to the energy.
f. Use $E=n^2h^2/8mL^2$ to find the exact value of the energy and compare the result with the answers to parts c and e. Take n=1.

Problem 4. An electron is confined to a box with sides at x=0 and x=1.6×10^{-8} m. There is an allowed value of the energy between E=9.3×10^{-22} J and E=9.5×10^{-22} J. Use the numerical technique outlined in problem 3 to find its value. Use $\Delta x=5\times10^{-11}$ m and take $\psi=0$, $d\psi/dx=2\times10^{12}$ m$^{-3/2}$ at x=0. Compare your answer with $n^2h^2/8mL^2$. Select the value of n which gives the closest agreement.

For a harmonic oscillator, the potential energy function is $\frac{1}{2}kx^2$, where k is a constant, and the Schrödinger equation can be written

$$d^2\psi/dx^2 = (8\pi^2m/h^2)(\tfrac{1}{2}kx^2 - E)\psi.\tag{27-11}$$

In the program of Fig. 27-1, the function f is evaluated using

$$(8\pi^2 m/h^2)*(.5*k*X3\uparrow 2-E)*X1\to X6$$

where values of m,h,k, and E must be supplied when the machine is programmed.

The particle is not confined by rigid walls and the wave function extends to all values of x. The boundary conditions are $\psi=0$ as x becomes either a large positive number or a large negative number.

If we were to follow the scheme used for the particle in a box, we would set $\psi=0$ for some large negative value of x, then integrate Eqn. 27-11 to find ψ for some large positive value of x. Finally, we would select E so that this last value of ψ is zero. This is not practical for two reasons. First, a large number of intervals must be used and this increases the running time enormously. Second, small errors in ψ and $d\psi/dx$ accumulate as the integration is carried out and eventually the error becomes quite large.

Another technique is needed. Since x enters the Schrödinger equation only as x^2 and as the independent variable in a second derivative, the equation is exactly the same if x is replaced by -x. This means that, if $\psi(x)$ is a solution, then $\psi(-x)$ is also. Furthermore, since there is only one probability function for a given E, it must be that either $\psi(-x)=\psi(x)$ or $\psi(-x)=-\psi(x)$. In the first case ψ is said to be an even function of x while, in the second case, it is said to be an odd function of x.

Look at the wave functions given in the table of section 26.1. The lowest energy state has an even wave function, with $\psi(-x)=\psi(x)$. The next state has an odd wave function, with $\psi(-x)=-\psi(x)$ and, in fact, the functions are alternatively even and odd.

If the function is even, then $d\psi/dx=0$ at x=0. We start the integration there and pick any convenient value for ψ. If the function is odd, then $\psi=0$ at x=0. Again we start the integration there and pick any convenient value for $d\psi/dx$.

There is also a problem with the boundary condition for large positive x. Again we do not wish to lose accuracy by using a large number of intervals. The problem, however, is more severe. For the harmonic oscillator, the wave function at large x is quite sensitive to the choice of E. For values a little different from an allowed value, the function is eventually proportional to $\exp(x^2/a^2)$ and quickly becomes very large.

To prevent this, the value of the energy must be selected with far more accuracy than we desire to know it or perhaps even more than the machine can handle. Much labor is required to obtain the required accuracy.

An approximate substitute boundary condition is needed. We pick a value of x which is small enough for the calculation to be handled, then select E so that ψ vanishes at that point. Linear interpolation is again used. If higher accuracy is desired, ψ can be forced to vanish at a larger value of x.

Problem 5.   An electron has a potential energy function given by $V=61x^2$ J, with x in meters. Use the program of Fig. 27-1 to find the energy and wave function for the lowest energy state. Take ψ to be 1×10^4 $m^{-\frac{1}{2}}$ and $d\psi/dx$ to be 0 at x=0. Use $\Delta x=2\times10^{-12}$ m and display values every 10 intervals until you obtain ψ at $x=3\times10^{-10}$ m.

a.   Take $E=5.5\times10^{-19}$ J.

b.   Take $E=6.5\times10^{-19}$ J.

c.   Use linear interpolation to estimate the value of E for which $\psi=0$ at $x=3\times10^{-10}$ m.

d.   Repeat the integration using the value of E found in part c.

e.   Finally, use the results of parts a and d to obtain a better estimate of the energy for which $\psi=0$ at $x=3\times10^{-10}$ m. Compare the estimate with $(h/4\pi)\omega_0$.

At $x=3\times10^{-10}$ m, the wave function oscillates from positive to negative as E increases from just below the particle energy to just above. If higher values of energy are considered, it would be found to continue this oscillation and, each time it passes through zero, E is close to an allowed energy value.

With the starting values of ψ and $d\psi/dx$ used in the last problem, only even wave functions can be found. In the next problem, the technique for finding the odd wave functions is used.

Problem 6.   An electron has a potential energy function given by $V=61x^2$ J, with x in meters. Use the program of Fig. 27-1 to find the energy and wave function for the next to lowest energy state. Take ψ to be 0 and $d\psi/dx$ to be 1×10^{16} $m^{-3/2}$ at x=0. Use $\Delta x=2\times10^{-12}$ m and display values every 10 intervals until you obtain ψ at $x=4\times10^{-10}$ m.

a.   Take $E=1.8\times10^{-18}$ J.

b.   Take $E=1.9\times10^{-18}$ J.

c. Use linear interpolation to estimate the value of E for which $\psi=0$ at $x=4\times 10^{-10}$ m.

d. Repeat the integration of Schrodinger's equation using the value of E found in part c.

e. Use the results of d and a to obtain a better estimate of the particle energy. Compare your answer with $(3h/4\pi)\omega_0$.

27.2 Square Well Problems

In this section, the Schrödinger equation is solved numerically for some situations which are variations on the particle in a box problem. It is worthwhile to work these problems in order to gain some insight into how the potential energy function influences the wave function and the probability distribution.

First, we consider solutions to the Schrödinger equation for regions in which the potential energy is constant. These solutions can be used to interpret solutions in more complicated cases.

The Schrödinger equation can be written in the form

$$d^2\psi/dx^2 = (8\pi^2 m/h^2)(V - E)\psi. \tag{27-12}$$

If V is constant and E>V, solutions have the form

$$\psi = A \sin\left[\sqrt{(8\pi^2 m/h^2)(E-V)}\ x + \phi\right] \tag{27-13}$$

where A and ϕ are constants. This expression can be checked by finding the second derivative with respect to x of the function and comparing it with the right side of Eqn. 27-12. The solutions oscillate and the wavelength of the oscillation is given by

$$\lambda = 2\pi/\sqrt{(8\pi^2 m/h^2)(E-V)}. \tag{27-14}$$

When the total and potential energies are nearly the same, the wavelength is long, but when the total energy is much greater than the potential energy, the wavelength is short. This behavior corresponds qualitatively to the reciprocal relationship between the particle momentum and the wavelength of the wave.

If V is constant and E<V, solutions have the form

$$\psi = A\ e^{\pm\sqrt{(8\pi^2 m/h^2)(V-E)}\ x}.$$

(27-15)

The wave function varies exponentially. In general, the wave function is a linear combination of the exponentially increasing function and the exponentially decreasing function. The wave function cannot increase indefinitely, however, since the probability must be less than 1. The application of boundary conditions is used to select the correct solution in a given situation.

If the potential energy is not constant, the solutions are not the ones given above but the correct solutions behave qualitatively in the same manner. Recall, for example, the wave functions for the harmonic oscillator. In the region between the classical turning points, the wave functions oscillate as functions of x. Wave functions for higher energy states exhibit more oscillations and the nodes are closer together. Beyond the classical turning points, the wave function goes to zero rapidly.

The wave function is a continuous function and, as the potential energy changes from place to place, the wave function changes in a smooth way. As the function goes from a region where E>V to a region where E<V, for example, the oscillatory function joins smoothly to a combination of exponential functions. The number of oscillations and the exponents of the exponential functions depend on the value of E and this must be chosen, for bound states, so that the wave function meets the boundary conditions.

These qualitative remarks are illustrated by the following examples. As you work the problems and plot the wave functions, use these remarks to help observe how the wave function is influenced by the potential energy function.

<u>Problem 1.</u>   The potential energy for an electron is given by $V(x)=4\times10^{-18}$ x between x=0 and $x=5\times10^{-7}$ m. At x=0 and at $x=5\times10^{-7}$ m there are rigid walls and the wave function vanishes. Here V is in joules and x is in meters. Use the program of Fig. 27-1, with $\Delta x=2.5\times10^{-9}$ m, to calculate $\psi(x)$. To find the energy, use a linear interpolation scheme, then integrate Schrödinger's equation for the energy obtained. Finally, carry out a second linear interpolation.

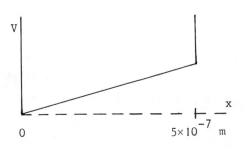

a. Find the wave function and energy level for the state with the lowest energy. As the first trial energy value, use $E=1\times10^{-24}$ J, then increment by 1×10^{-25} J. Do not normalize the wave function.

b. Find the wave function and energy level for the state with the lowest energy greater than 3.1×10^{-24} J. Start with $E=3.1\times10^{-24}$ J, then increment by 1×10^{-25} J. Do not normalize the wave function.

c. Draw graphs of the two probability densities. Plot points every 5×10^{-8} m.

For both wave functions, the influence of the increasing potential energy is evident. In part b, the energy is greater than the potential energy at all points inside the box and the probability density oscillates. It has 5 nodes, counting the end points, just like the n=4 wave function for the case V=0 inside the box. Careful observation reveals, however, that the nodes are further apart at the right end of the box where the potential energy is higher.

The function of part a starts out, near x=0, like the n=1 wave function for a particle in a box with V=0. The rising potential energy causes the peak to be in the left side of the box, however, rather than in the center. Then the classical turning point is reached, near $x=2.7\times10^{-7}$ m, and the wave function decreases rapidly thereafter. It is no longer oscillatory.

Problem 2. An electron is in a box with a rigid wall at x=0. For $x<x_1$, the potential energy vanishes but, at $x=x_1$, it jumps to V_0.

a. Take $V_0=4.7\times10^{-24}$ J, $x_1=5\times10^{-7}$ m, and find the wave functions and energy levels for the states with the two lowest energies. Use $\Delta x=2.5\times10^{-9}$ m. The wave function is exponential beyond $x=x_1$ and you must use a technique similar to that used for harmonic oscillator functions. Pick the energy level by requiring the wave function to vanish

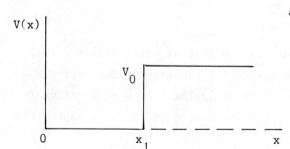

at $x=7\times10^{-7}$ m. For the lowest energy state, use $E=2.1\times10^{-25}$ J as the first trial energy value, then increment by 1×10^{-26} J. For the second state, start with $E=8.3\times10^{-25}$ J, then increment by 1×10^{-26} J. Use linear interpolation to estimate the value of E for which $\psi=0$ at $x=7\times10^{-7}$ m, then integrate the Schrödinger equation for that value of E. Carry out a second linear interpolation. The instructions

552

for calculating f in the program of Fig. 27-1 must include a transfer statement so that $f=-(8\pi^2 m/h^2)E\psi$ is used for $x<x_1$ and $f=(8\pi^2 m/h^2)(V_0-E)\psi$ is used for $x>x_1$.

b. For each of the wave functions, draw a graph of the probability density as a function of x. Plot points every 5×10^{-8} m and mark the position of x_1 on the graphs.

For the low energy states, the wave functions are quite similar to the wave functions for a particle in a box with rigid walls at $x=0$ and $x=x_1$. But there are some important differences.

Unlike the case of the box with rigid walls, the wave functions now extend into the region for which V>E. There they decrease exponentially toward 0.

The energy of the lowest state is slightly lower than the lowest energy for a particle in a box with rigid walls. That this is so is intimately connected with the extension of the wave function beyond $x=x_1$. For the box with rigid walls, the wave function is half a cycle of a sine function. It fits in the box in such a way that it vanishes at both walls. Now the function can be made to extend beyond x_1 and decrease exponentially by expanding the sine function slightly so that less than half a cycle is between $x=0$ and $x=x_1$. If ψ is positive in the box, then it is positive at $x=x_1$ and has negative slope. The exponential which joins smoothly to it then continues the decrease toward zero. In order to expand the wave function, the energy is decreased.

You should convince yourself that the boundary condition cannot be met if the sine function is contracted a bit so that slightly more than half a cycle lies between $x=0$ and $x=x_1$.

If the energy is increased, the sine function contracts, until almost a full cycle lies between $x=0$ and $x=x_1$. For a certain value of the energy, a decreasing exponential function can be joined smoothly to it. This is then the wave function for the second level. The qualitative description can be continued in the same manner for the higher levels.

If the wave function were normalized, the graphs would show that the function for the second level extends a greater distance beyond x_1 than does the function for the lowest level. This trend also continues for higher levels.

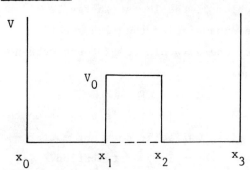

Problem 3. The potential energy function for an electron is as shown, with $V_0 = 2.3 \times 10^{-24}$ J, $x_0 = 0$, $x_1 = 2 \times 10^{-7}$ m, $x_2 = 3 \times 10^{-7}$ m, and $x_3 = 5 \times 10^{-7}$ m. The walls at x_0 and x_3 are rigid. The instructions for calculating f in the program of Fig. 27-1 must contain some conditional transfer statements so that $f = -(8\pi^2 m/h^2)E\psi$ is used for $x_0 < x < x_1$ and $x_2 < x < x_3$, and $f = (8\pi^2 m/h^2)(V_0 - E)\psi$ is used for $x_1 < x < x_2$. For the integrations, use $\Delta x = 2.5 \times 10^{-9}$ m.

a. Find the wave function and energy level for the state with the lowest energy. Start with $E = 8 \times 10^{-25}$ J and increment by 2×10^{-26} J.

b. Find the wave function and energy level for the state with the next highest energy. Start with $E = 1 \times 10^{-24}$ J and increment by 2×10^{-25} J.

c. For each of the two states, plot the unnormalized probability density as a function of x. Plot points every 5×10^{-8} m and mark the points x_0, x_1, x_2, and x_3 on the graph.

In the region of the potential energy bump, the wave function is a combination of both the exponentially increasing and the exponentially decreasing solutions to the Schrodinger equation. To the left of the bump, the wave function is oscillatory and, if it starts out with positive slope, then near x_1 it has negative slope. Just beyond x_1, it is predominantly the decreasing function which joins smoothly to it. As x increases, the exponentially increasing function becomes more prominent. To the right of the bump, the wave function is again oscillatory.

For the low lying levels considered in this problem, it is instructive to compare the wave function with those for two separated boxes. The first box extends from x=0 to $x = x_1$ and the second box extends from $x = x_2$ to $x = x_3$.

For the lowest state, the wave function on each side is quite similar to that for a box with rigid walls. Since the wall on one side is not rigid, a little less than half a cycle of the sine function is inside the box. This function then joins smoothly to a combination of exponential functions. For the wave function on the left, the combination decreases toward the right and, for the wave function on the right, the combination decreases toward the left. In the region between the boxes, these two functions join smoothly to each other.

The lowest energy level for an electron in a 2×10^{-7} m wide box with rigid walls is about 1.5×10^{-24} J. For the situation given in problem 3, the lowest energy level is about 8.1×10^{-25} J. The lower energy accounts for the expansion of the sine function inside the box. This energy is larger than the lowest energy would be if there were no potential energy bump in the middle of the original box.

The state considered in part b shows another feature. If $\psi(x)$ is a solution to Schrödinger's equation, so is $-\psi(x)$. For a box with rigid walls, it does not matter which solution is used since they both produce the same probability distribution. If, however, the situation of problem 3 is thought of as two boxes, the wave function can be constructed by adding or subtracting the functions appropriate for the individual boxes. The lowest state is formed by adding them. The next state is formed by subtracting them. Notice, in part b, that on each side, the function looks similar to the function for the lowest state for a box with rigid walls, but on one side it is positive while on the other side it is negative. The energy is again lower than the lowest energy allowed for one of the narrow boxes and the sine functions are expanded somewhat.

The ideas demonstrated by this simple example are useful in the study of much more complicated wave functions. In a molecule or crystal, the wave functions for the low lying states behave in much the same manner. There the potential energy function is found by considering the electrostatic attraction of the nuclei and the repulsion of the other electrons. It is quite complicated. Near any nucleus, however, the wave function is very nearly the same as it is for the atom formed by that nucleus and its complement of electrons. These wave functions, one centered at each nucleus, then join smoothly in the region between nuclei. The energy levels are different for the molecule or crystal and this difference allows the smooth joining of the functions.

<u>Problem 4.</u>   The potential energy function for an electron is shown to the left.

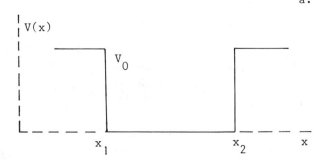

a.   Take $V_0 = 4.7 \times 10^{-24}$ J, $x_1 = 0$, $x_2 = 5 \times 10^{-7}$ m, and find the wave functions and energy levels for the states with the two lowest energies. Use $\Delta x = 2 \times 10^{-9}$ m. The wave functions decrease exponentially in magnitude for $x < x_1$ and $x > x_2$. They are either symmetric or antisymmetric about $x = (x_2 + x_1)/2$. Start the integration at that point either with $\psi = 1 \times 10^5$ m$^{-\frac{1}{2}}$, $d\psi/dx = 0$ or with $\psi = 0$,

555

$d\psi/dx = 1 \times 10^{12}$ m$^{-3/2}$. Approximate the boundary condition by requiring the wave function to vanish at $x = 7 \times 10^{-7}$ m. For the lowest state, take the first trial energy value to be 1.8×10^{-25} J, then increment by 1×10^{-26} J. For the second state, take the first trial energy value to be 7.3×10^{-25} J, then increment by 1×10^{-26} J. Use linear interpolation to estimate the value of E for which $\psi = 0$ at $x = 7 \times 10^{-7}$ m, then integrate Schrödinger's equation using that value of E.

b.   For each state, draw a graph of the probability density as a function of x. Plot points every 5×10^{-8} m. Mark the positions of x_1 and x_2 on the graphs. Do not normalize the wave functions.

It has been observed that, for many of the situations considered, the wave function extends into regions for which the potential energy is greater than the total energy of the particle. It is only when the potential energy becomes infinite, as for the box with rigid walls, that this extension of the wave function does not occur. In the situations considered so far, we applied the boundary condition that the wave function must eventually approach zero inside the classically forbidden region.

We now consider another case. A typical potential energy function is shown in Fig. 27-2. As before, there is a classically forbidden region but it is narrow.

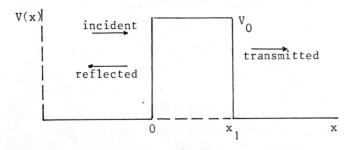

Figure 27-2. A potential energy barrier. The directions of travel for the incident, reflected, and transmitted waves are shown.

The wave function does not decrease to zero before the edge of the region and there is some probability that the particle can be found on the far side of the classically forbidden region.

A plane wave is incident from the left on the barrier. The energy of the particle is less than V_0. There is a reflected wave in the region to the left and a transmitted wave in the region to the right. The square of the amplitude of the transmitted wave gives the probability that the particle can be found to the right of the barrier. If it is, it is said to have tunneled through the barrier.

In the region to the right of the barrier, the wave function has the form

$$\Psi(x,t) = A \, e^{i(kx-\omega t)} \qquad \text{for } x>x_1, \qquad (27\text{-}16)$$

where $k=2\pi p/h$ and $\omega=2\pi E/h$. Here p is the momentum of the particle and E is its energy. Since the particle is free, the energy is not quantized and it can have any value. A is a constant.

In the region to the left of the barrier, the wave function has the form

$$\Psi(x,t) = B \, e^{i(kx-\omega t)} + C \, e^{i(-kx-\omega t)} \qquad \text{for } x<0. \qquad (27\text{-}17)$$

Here B and C are constants. The first term represents the right going incident wave, while the second term represents the left going reflected wave.

In the barrier, the wave function is a linear combination of exponentially increasing and exponentially decreasing functions.

Usually it is the incident wave which is known and the reflected and transmitted waves are obtained from the Schrodinger equation. This procedure cannot be followed when numerical techniques are used. Instead, we suppose the wave function and its derivative are known for some point to the right of the barrier, then use the Schrödinger equation to find the function for points to the left. This means the interval Δx is a negative number.

Since the time dependent factor $e^{-i\omega t}$ is the same at all points, we can omit writing it and deal only with the part of the wave function that depends on x. The problem then is: given $\psi(x)=Ae^{ikx}$ for $x>x_1$, integrate the Schrödinger equation to find $\psi(x)$ for $x<x_1$.

We must deal with a complex wave function. This causes no concern since both the real and imaginary parts satisfy Schrödinger's equation, independently of each other. What links them is the boundary condition. To the right of the barrier

$$\psi(x) = A \cos(kx) + iA \sin(kx) \qquad (27\text{-}18)$$

and

$$d\psi/dx = -kA \sin(kx) + ikA \cos(kx). \qquad (27\text{-}19)$$

The Schrodinger equation is integrated twice, once for the real part and once for the imaginary part. To simplify the procedure, pick the starting place where $\cos(kx)=+1$ and $\sin(kx)=0$. The real part of the wave function is found by placing $\psi=A$ and $d\psi/dx=0$ at that point, then integrating the Schrödinger equation. The imaginary part is found by placing $\psi=0$ and $d\psi/dx=kA$ at that point, then integrating the Schrödinger equation again.

Once the real and imaginary parts of ψ have been found, the probability density $|\psi|^2$ is calculated for a given point by using

$$|\psi(x)|^2 = \psi_R^2(x) + \psi_I^2(x). \qquad (27\text{-}20)$$

Problem 5. A potential barrier is as shown in Fig. 27-2. An electron with energy $E=1\times10^{-24}$ J impinges on it from the left. Take $V_0=1.15\times10^{-24}$ J and $x_1=2\times10^{-7}$ m.

a. Calculate the momentum of the electron, then calculate the propagation constant k for the electron wave.

b. Suppose the real part of the wave function is 1×10^4 m$^{-\frac{1}{2}}$ and its derivative is 0, at $x=7\times10^{-7}$ m. Take $\Delta x=-5\times10^{-9}$ m and calculate the real part of the wave function for points 5×10^{-8} m apart, to the left of the starting point. Stop when $x=-2\times10^{-7}$ m.

c. Calculate the imaginary part of the wave function and its derivative for $x=7\times10^{-7}$ m.

d. Take $\Delta x=-5\times10^{-9}$ m and calculate the imaginary part of the wave function for the points used in part b.

e. For each of the coordinate points of parts b and d, calculate the unnormalized probability density and plot it as a function of x. Mark the positions of the barrier edges on the graph.

Look at the graph you drew for part e. To the left of the barrier, the probability density oscillates with x. This is because the wave function there is a linear combination of the incident and reflected waves. It is quite large at places. As the incident wave enters the barrier its amplitude decreases exponentially and this decrease is evident on the probability density graph. To the right of the barrier, there is a single plane wave, traveling to the right and the probability density is constant.

The amplitude of the incident wave can be found. It is of interest since, if it is known, it is possible to calculate the chance that the particle penetrates the barrier. We assume B and C are real and positive. If they are not, the analysis produces the same result but the mathematics is more complicated. On the left side of the barrier

$$|\psi|^2 = B^2 + C^2 + 2BC \cos(2kx), \qquad (27\text{-}21)$$

a result which follows easily from Eqn. 27-17 and the identity $2\cos(2kx)=\exp(i2kx) + \exp(-i2kx)$. The maximum value of $|\psi|^2$ occurs where $\cos(2kx)=1$ and it is

$$|\psi|^2_{max} = (B + C)^2. \qquad (27\text{-}22)$$

The minimum value of $|\psi|^2$ occurs where $\cos(2kx)=-1$ and it is

$$|\psi|^2_{min} = (B - C)^2. \qquad (27\text{-}23)$$

When these two equations are solved for B, the result is

$$B = \tfrac{1}{2}(|\psi|_{max} + |\psi|_{min}). \qquad (27\text{-}24)$$

The probability that the particle penetrates the barrier is given by $(A/B)^2$.

In problem 5, the points were taken too far apart to obtain accurate values for $|\psi|^2_{max}$ and $|\psi|^2_{min}$, but rough estimates yield 0.25 for the probability of tunneling. You should verify this result.

In problem 3, the particle energy was only slightly below the top of the barrier. If the energy is less, the probability of tunneling is less. The probability of tunneling is also less if the barrier is wider. In the following problems, these assertions are demonstrated.

Problem 6.   For the potential barrier of Fig. 27-2, take $V_0=1.15\times10^{-24}$ J and $x_1=2\times10^{-7}$ m. An electron is incident from the left, with energy $E=9\times10^{-25}$ J.

a.   Calculate the momentum of the electron and the propagation constant of its wave.

b.   Take the real part of the wave function to be 1×10^4 m$^{-\frac{1}{2}}$ and its derivative to be 0 at $x=2.5\times10^{-7}$ m. Calculate the imaginary part of the wave function and its derivative at that point. Then use $\Delta x=-5\times10^{-9}$ m to calculate the real and imaginary parts for points 5×10^{-8} m apart, to the left of the starting point. Stop at $x = -2 \times 10^{-7}$ m.

c. For each of the points in part b, calculate the unnormalized probability density and plot it as a function of x. Mark the positions of the barrier edges on the graph.

d. Estimate the amplitude of the incident wave and calculate the probability that the particle tunnels through the barrier.

Problem 7.   The potential barrier is as shown in Fig. 27-2 except that the left edge is at $x = -5 \times 10^{-8}$ m. Take $V_0 = 1.15 \times 10^{-24}$ J and $x_1 = 2 \times 10^{-7}$ m. An electron with energy $E = 1 \times 10^{-24}$ J is incident from the left.

a. Take the real part of the wave function to be 1×10^4 $m^{-\frac{1}{2}}$ and its derivative to be 0 at $x = 2.5 \times 10^{-7}$ m. Calculate the imaginary part of the wave function and its derivative at that point. Then use $\Delta x = -5 \times 10^{-9}$ m to calculate the real and imaginary parts for points 5×10^{-8} m apart, to the left of the starting point. Stop when $x = -2 \times 10^{-7}$ m. The propagation constant was computed in answer to problem 5.

b. For each of the points in part a, calculate the unnormalized probability density and plot it as a function of x. Mark the positions of the barrier edges on the graph.

c. Estimate the amplitude of the incident wave and calculate the probability that the particle tunnels through the barrier.

One model of α decay makes use of the idea of tunneling through a barrier. An α particle is a helium nucleus, formed by two protons and two neutrons, tightly bound together. An α particle may be formed inside some heavy nucleus by the particles there. If formed, it is strongly attracted to the other particles in the nucleus and these attractive forces form a high potential energy barrier. The attractive forces, however, have a short range, on the order of 1×10^{-15} m and, outside this range, the electrostatic repulsion of the protons left behind in the nucleus causes the potential energy function to decrease. The α particle has a small probability of tunneling through the barrier and, when it does, a radioactive decay process is said to occur. The barrier for α particles is much more complicated than the one we considered in the preceding problems but the idea of barrier tunneling is the same.

Chapter 28

TOPICS IN SOLID STATE AND NUCLEAR PHYSICS

Programs developed previously are used to investigate two topics in solid state
and nuclear physics. The distribution of electrons among conduction band, valence
band, and impurity states of a semiconductor are studied. The temperature dependence
of these populations and the role played by doping are emphasized. This leads to a
calculation of the contact potential for a p-n junction. In nuclear physics, the
radioactive decay law is studied and solutions for various physical situations are
obtained numerically. Techniques for the analysis of experimental data are presented.
This chapter is intended for use in association with Chapters 46 and 47 of FUNDAMENTALS
OF PHYSICS (EXTENDED).

28.1   Electrons in Semiconductors

Electrons in crystalline solids are bound and their energies are quantized.
Fig. 28-1 is a schematic energy level diagram for electrons in a typical solid. The
allowed energy levels are bunched together in groups, called bands. The number of individual states in a band is proportional to the number of atoms in the crystal and, for macroscopic samples, is on the order of 10^{15} to 10^{22}. Levels within a band are so close together they cannot be shown individually on the diagram. Although, for some crystals, two or more bands may overlap, they are generally separated by gaps. In a pure crystal, no electrons can have an energy which is in one of the gaps.

Low lying bands are very narrow. An electron in one of them is tightly bound to a single atom. Its wave function and energy are almost identical to the wave function and energy, respectively, of the corresponding electron in a single, isolated atom of the same type. In the diagram, there are a large number of states in

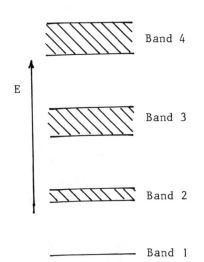

Figure 28-1.   Schematic energy
level diagram for electrons in
a solid.

560

band 1 but they all have nearly the same energy associated with them.

At higher energy, the bands are wider and the electrons in these states have wave functions which extend throughout the crystal. The most energetic electrons are responsible for most of the electrical and optical properties of the material and these are the electrons of interest in this section.

Electrons obey the Pauli exclusion principle: while an electron occupies any state, no other electron can be in that state. This principle is quite general and holds for all systems containing more than one electron. Protons, neutrons, and certain other particles also obey exclusion principles but photons and mesons do not. This means that two or more photons, for example, can be in the same state, while two electrons cannot be in the same state.

The lowest total energy for the collection of electrons in a crystal cannot be achieved by placing all of the electrons in the lowest energy state of the lowest band. This configuration is in violation of the exclusion principle.

Instead, the lowest possible total energy for a collection of N electrons is achieved when the N states with the lowest energies are occupied and states with higher energy are vacant. This configuration of lowest total energy can be found, in principle, by listing the states in order of increasing energy, then counting the states. The highest filled state is reached when the number of states counted equals the number of electrons in the crystal.

The electrons need not be in the configuration with lowest total energy. When energy is supplied to an electron, it makes a transition from its original state to a state at higher energy. The state it enters must be empty originally.

We have already learned about one method for exciting electrons from one state to another. A photon, on entering the material, may interact with an electron and transfer some or all of its energy to the electron in a Compton or photoelectric interaction. When we investigated these effects earlier we implicitly assumed the availability of an empty state with energy equal to the value we calculated for the electron after the interaction.

In this section we are concerned with another way in which electrons might be promoted from lower states to higher states. That is by heating the material. The

562

process is called thermal excitation of electrons. In the heating process, energy is supplied primarily to the atomic nuclei in the material and, because the nuclei and electrons interact electrically with each other, some electrons receive energy and are promoted to states with higher energy. At higher temperatures electrons occupy higher states than at lower temperatures, and it is chiefly electrons from the highest filled states which are promoted to the lowest empty states.

The distribution of electrons among the various states plays an important role in determining the electrical current in the material. In the absence of an external electric field the current is zero because, for every electron going in any direction, there is another electron going with the same speed in the opposite direction.

For purposes of calculating the electric current, a velocity can be associated with each state and, for a given state and its associated velocity, the geometrical symmetry of the crystal guarantees there is another state in the same band with velocity in the opposite direction. In thermal equilibrium with no external field present, the two states with oppositely directed velocities have equal probability of being occupied.

When an electric field is turned on, it causes electrons to jump to states, within the same band, for which the velocity is predominantly in the direction opposite to that of the field. This empties some states for which the velocity is predominantly in the same direction as the field and the net result is a collection of electrons with average velocity opposite to the field. Since electrons carry negative charge, this constitutes a current in the direction of the field.

In order for an electric field to create a current there must be empty states in the band. The electrons cannot jump to states which are already occupied. A completely filled band cannot contribute to the electric current.

For many materials, the number of electrons in the crystal is not equal to the number of states in an integer number of bands. Even in the lowest energy configuration the highest occupied band is only partially filled and there are many empty states, in that band, which can receive electrons. For a given applied electric field, a large number of electrons make transitions and contribute to the current. These materials are metals and are good conductors of electricity.

For a great many crystals, the number of electrons is just right to completely fill an integer number of bands with no electrons left over. If the bands do not overlap, these materials are either insulators or semiconductors. If the electrons are in the lowest energy configuration, there can be no electric current since all of the states in all occupied bands are filled. There are no bands which are only partially filled.

At the absolute zero of temperature, the crystal is in the lowest energy configuration. For insulators and semiconductors, the highest band which is filled at $T=0^{\circ}$ K is called the valence band and the lowest band which is empty at $T=0^{\circ}$ K is called the conduction band.

As the crystal is heated and its temperature raised, some electrons are promoted across the gap between the valence and conduction bands. To achieve higher temperatures, more energy must be supplied to the crystal and the electrons receive a share. More electrons are promoted across the gap and states with higher energies become occupied.

At temperatures above absolute zero, neither the valence nor the conduction band is completely filled and electrons in both can contribute to the electric current when an electric field is turned on.

Semiconductors and insulators differ in the width of the gap between the valence and conduction bands. For insulators the gap is wide, on the order of several electron volts, and extremely few electrons are promoted even at temperatures near the melting point of the crystal. The electric current is small for any applied field. For semiconductors the gap is narrow, about 1 electron volt or less, and considerably more electrons are promoted across it. With the same applied electric field the current is larger for semiconductors than for insulators. At room temperature, a semiconductor has enough electrons in partially filled bands to produce a measurable current when an electric field is turned on but its conductivity is not as great as that of a typical metal.

For a typical semiconductor at room temperature, the number of electrons excited to the conduction band is larger than for an insulator but it is still a very small fraction of the number of states available in the band. Likewise, the number of empty states left behind in the valence band is a very small fraction of the number of states in that band. It is convenient to deal with the small number of empty states in the valence band rather than with the large number of electrons in that band. The empty states are assigned properties just as if they were occupied by particles. The fictitious

564

particles are called holes and properties such as momentum, energy, and velocity are assigned them so that, in calculations of crystalline properties, the holes can be used rather than the electrons.

In many respects, a hole acts like a particle with a positive charge, equal in magnitude to the charge on the electron. When an electric field is turned on, there is a net motion of holes in the direction of the field. They contribute to the current and, since they act like positively charged particles, their contribution is in the direction of the field. The total current in a semiconductor is made up of two parts: moving electrons in the conduction band and moving holes in the valence band.

The basis for attributing the characteristics of a positive charge to a hole lies in the pairing of the electrons. Recall that a filled band does not contribute to the current because the electrons can be paired in such a way that the two members of any pair have velocities which are equal in magnitude and opposite in direction. A band with one vacant state can be considered to be a filled band plus a positive charge, to cancel one of the electrons. The filled band does not contribute to the current but the fictitious positive charge does. The model can be carried further. When an electric field is turned on, the change in the total momentum for the band is predominantly in the direction of the field and the collection of electrons behaves like a single positive charge, which accelerates in the direction of the field.

In this section, we shall investigate the density of electrons in the conduction band and the density of holes in the valence band, both as functions of the temperature. We suppose there are N(E) states per unit volume with energy E and we need to know how many of these states are occupied at temperature T. An expression for this quantity can be derived using the principles of thermodynamics, with the result that the number of electrons per unit volume with energy E is given by

$$n(E) = \frac{N(E)}{e^{(E-E_F)/kT}+1} \qquad (28-1)$$

where k is Boltzmann's constant (k=1.380662×10^{-23} J/$^{\circ}$K) and E_F is a quantity called the Fermi energy. It will be discussed below. The temperature is in degrees Kelvin above the absolute zero. Note that, for E in a gap, N(E)=0 and Eqn. 28-1 predicts that the density of electrons with energy in a gap vanishes, as it should.

Eqn. 28-1 gives the density of electrons with energy E when the crystal is in thermodynamic equilibrium at temperature T. In thermodynamic equilibrium electrons are continually falling to lower levels and are being promoted to higher levels by means of interactions with the nuclei. On the average, however, there is no net energy transfer between the electron system and the nuclei. Except for small fluctuations, which we neglect, the density of electrons with any energy E remains constant in time. As Eqn. 28-1 indicates, it can be changed by changing the temperature.

The quantity $1/\left[e^{(E-E_F)/kT} + 1\right]$ can be regarded as the probability that a state with energy E is occupied by an electron when the crystal has temperature T. It gives the fraction of time during which a state with energy E is occupied.

The Fermi energy E_F is a parameter which is characteristic of the material at the given temperature. It must be calculated for each different material and for each temperature of interest. The condition used for its calculation is that when the electron densities, as predicted by Eqn. 28-1 for all the electron energy levels, are summed, the result must be the overall electron density for the crystal as determined by experiment or other means. E_F is not necessarily any one of the allowed energy levels for an electron in the crystal. In the rest of this section there are several examples of calculations of the Fermi energy.

The Fermi energy is extremely important for the determination of the density of electrons in the conduction band and the density of holes in the valence band. As Eqn. 28-1 indicates, these densities are quite sensitive to the position of the Fermi energy relative to the electron energy levels.

For pure semiconductors at $T=0^{o}$ K the Fermi energy is at the midpoint of the gap between the valence and conduction bands. As the temperature is increased, E_F moves from this position but, for most semiconductors, it remains within the gap.

The electrical properties of semiconductors can be altered significantly by adding certain impurities to the material. When this is done the position of the Fermi energy is changed. The distribution of the electrons among the various states is also changed and, from a calculational point of view, the distribution of electrons is controlled by the position of the Fermi level. Several examples of Fermi energy calculations for materials with impurities are given later.

The Pauli exclusion principle is incorporated into the function $1/\left[e^{(E-E_F)/kT} + 1\right]$. For no value of E is it greater than 1 and the number of electrons with energy E is never greater than the number of states with that energy.

For electrons in the valence band and lower bands, $E-E_F$ is negative and $e^{(E-E_F)/kT}$ is a positive number less than 1. If $(E-E_F)/kT$ is less than -7, the exponential is less than 1×10^{-3} and the denominator has a value close to 1. In this case, $n(E) \simeq N(E)$ and these levels are almost completely filled. Near the top of the valence band, there are a few states for which $n(E)$ is slightly less than $N(E)$. There are empty states and these are the holes.

For electrons in the conduction band and higher bands, $E-E_F$ is positive and $e^{(E-E_F)}$ is a positive number greater than 1. If $(E-E_F)$ is greater than +7, the exponential is greater than 1×10^3 and $n(E) \simeq N(E) e^{-(E-E_F)/kT}$. The density of electrons in these states is extremely small but these electrons are important for electrical conduction.

The program of Fig. 2-4 can be used to plot $1/\left[e^{(E-E_F)/kT} + 1\right]$ as a function of E. Enter E_F into X4 and T into X5. The initial value of E is entered into X1, the final value into X2, and the increment into X3. The instruction at line 120 should be

$$1/(\text{EXP}((X1-X4)/(1.380662E-23*X5))+1) \rightarrow X6$$

The program can also be used to plot $n(E)$. Simply replace the first 1 in the above instruction by instructions for the calculation of $N(E)$.

Some of the problems ask for the value of the function for a single value of E, rather than for a sequence of values. The program can again be used. Simply enter the value into both X1 and X2.

Problem 1. For germanium, the gap between the valence and conduction bands is about 1.1×10^{-19} J. Take the top of the valence band to be $E_v = 0$, the Fermi energy to be 5.5×10^{-20} J (midway in the gap), and the temperature to be 300° K. As we shall see later, this is not quite the correct value for the Fermi energy.

a. Plot $1/\left[e^{(E-E_F)/kT} + 1\right]$ as a function of the energy from $E=-6 \times 10^{-20}$ J to $E=+1.2 \times 10^{-19}$ J. Plot points every 1×10^{-20} J. This function gives the fraction of states which are occupied for each energy. On the graph, mark the positions of the top of the valence band and the bottom of the conduction band.

b. Calculate the fraction of occupied states at the top of the valence band and at the bottom of the conduction band.

c. Calculate the fraction of states with holes at the top of the valence band. The fraction of occupied states and the fraction of unoccupied states must sum to 1.

d. Calculate the fraction of occupied states 3×10^{-20} J above the bottom of the conduction band.

As you discovered, the fraction of occupied states at the bottom of the conduction band is small and it decreases rapidly for higher energies. Similarly, the fraction of unoccupied states at the top of the valence band is small and it decreases rapidly for lower energies.

In order to simplify calculations, it is often assumed that all of the electrons in the conduction band have the same energy E_c and all of the holes in the valence band have the same energy E_v. These are taken to be the bottom of the conduction band and the top of the valence band, respectively.

As can be seen from the above calculation, there is a loss in the number of significant figures if the fraction of unoccupied states in the valence band is calculated by first computing the fraction of occupied states, then subtracting the result from 1. It is better to carry out the subtraction algebraically. The steps are

$$1 - \frac{1}{e^{(E-E_F)/kT} + 1} = \frac{e^{(E-E_F)/kT} + 1 - 1}{e^{(E-E_F)/kT} + 1}$$

$$= \frac{e^{(E-E_F)/kT}}{e^{(E-E_F)/kT} + 1} = \frac{1}{e^{-(E-E_F)/kT} + 1} . \qquad (28\text{-}2)$$

In the first step, the two terms were written with a common denominator, while in the last step, both numerator and denominator were multiplied by $e^{-(E-E_F)/kT}$ and the order of terms in the denominator was reversed.

The density p of holes at energy E is given by

$$p = \frac{N(E)}{e^{-(E-E_F)/kT} + 1}, \qquad (28\text{-}3)$$

where N(E) is the number of states per unit volume with energy E. This expression is quite similar to the expression for the density of electrons. The sign of the exponent, however, is different and that produces a large numerical difference.

The program of Fig. 2-4 can be modified to calculate both the fraction of occupied states and the fraction of unoccupied states for energy E. Retain the modification given just prior to problem 1 and add the instruction

$$1/(EXP((X4-X1)/(1.380662E-23*X5))+1) \rightarrow X7$$

to line 120, then have the machine display both X6 and X7.

As the temperature is raised, the fraction of occupied states in the conduction band and the fraction of unoccupied states in the valence band both increase. Electrons are thermally excited across the gap, leaving behind a larger number of empty states.

Problem 2.   For germanium at T=1000° K, calculate the fraction of occupied states at the bottom of the conduction band and the fraction of unoccupied states at the top of the valence band. Take the gap to be 1.1×10^{-19} J wide and assume the Fermi energy is at the center of the gap. This assumption is not precisely valid, as we shall see.

We must now take into account the shift in the Fermi energy as the temperature changes. You can visualize what happens when the Fermi energy changes by looking at the graph you drew in response to problem 1. If the Fermi energy increases, the whole curve moves toward higher energy, leaving behind the location of the electron energy levels. It is clear that there are more electrons in the conduction band and fewer holes in the valence band. When E_F is increased, Eqn. 28-1 predicts a larger number of electrons in the crystal. There are additional electrons in both the valence and conduction bands.

If the Fermi energy decreases, Eqn. 28-1 predicts that the total number of electrons in these bands decreases.

Problem 3.   For germanium at 300° K, calculate the fraction of occupied states at the bottom of the conduction band and the fraction of unoccupied states at the top of the valence band. Take the gap to be 1.1×10^{-19} J wide and consider each of the following possible values of the Fermi energy.

a. $E_F = 1 \times 10^{-20}$ J above the top of the valence band.

b. $E_F = 2.5 \times 10^{-20}$ J above the top of the valence band.

c. $E_F = 2.5 \times 10^{-20}$ J below the bottom of the conduction band.

d. $E_F = 1 \times 10^{-20}$ J below the bottom of the conduction band.

For the first two values of the Fermi energy, there are fewer electrons in the conduction band and more holes in the valence band than for the value considered in problem 1. For the last two values, there are more electrons and fewer holes.

The symmetry of the function $1/\left[e^{(E-E_F)/kT} + 1\right]$ is evident in the results. The fraction of states ΔE above the Fermi energy which are occupied is exactly the same as the fraction of states ΔE below the Fermi energy which are empty.

We are now in a position to calculate the Fermi energy for the crystal. At $T = 0^\circ$ K, all the valence band states are occupied and all the conduction band states are empty. The total number of electrons in these two bands is exactly equal to the number of states in the valence band. At higher temperatures, the sum of the number of electrons in the conduction band and the number of electrons in the valence band must again equal the number of states in the valence band. This condition is exactly the same as the condition that the number of electrons in the conduction band is equal to the number of holes in the valence band and it is the condition used to calculate the Fermi energy. E_F is adjusted so that Eqn. 28-1 predicts the correct density of electrons for the crystal.

We use the two level model of a semiconductor. We consider N_c states per unit volume in the conduction band, all with energy E_c. Likewise, we consider N_v states per unit volume in the valence band, all with energy E_v. Only states near the gap are considered, not all of the states in the two bands. Then the density of electrons in the conduction band is given by

$$n = \frac{N_c}{e^{(E_c - E_F)/kT} + 1} \qquad (28\text{-}4)$$

and the density of holes in the valence band is given by

$$p = \frac{N_v}{e^{-(E_v - E_F)/kT} + 1} \, . \qquad (28\text{-}5)$$

The Fermi energy is adjusted so that n=p, or

$$\frac{N_c}{e^{(E_c-E_F)/kT} + 1} - \frac{N_v}{e^{-(E_v-E_F)/kT} + 1} = 0. \qquad (28\text{-}6)$$

If $(E_c-E_F)/kT \gg 1$ and $(E_F-E_v)/kT \gg 1$, as they usually are, this equation can be solved analytically for E_F. The result is

$$E_F = \tfrac{1}{2}(E_c+E_v) + (kT/2)\ \ell n(N_v/N_c). \qquad (28\text{-}7)$$

To obtain this result, the 1's which appear in the denominators of Eqn. 28-6 are neglected. Note that Eqn. 28-7 predicts that the Fermi energy is at the center of the gap for $T=0^{\circ}$ K and that it shifts away from the center as the temperature increases.

The exact expression, Eqn. 28-6, can be solved numerically for E_F by using one of the root finding programs. The variable is E_F and, if the binary search program is used, values which straddle the root are placed in X8 and X9 respectively. Until you gain some experience with variations in the Fermi energy, try E_v as the lower limit and E_c as the upper limit of the search.

At line 510 of the binary search program, the instruction should be

$(N_c)/(EXP((E_c-X8)/(kT))+1)-(N_v)/(EXP((X8-E_v)/(kT))+1) \rightarrow X2$

At lines 520 and 600 the instructions are the same except X8 is replaced by X9 or X4 and X2 is replaced by X3 or X5, as appropriate. Values for N_c, N_v, E_c, E_v, and kT must be supplied at the time of programming. If your machine has sufficient memory, it is worthwhile to have the machine store these values, then recall them as needed.

Once the Fermi energy has been found, the density of electrons in the conduction band and the density of holes in the valence band can be calculated using Eqns. 28-4 and 28-5, respectively. Appropriate instructions can be added after line 550 of the binary search program, if desired.

To make the two level model more accurate, N_c and N_v are usually taken to be temperature dependent. This modification corrects the model to take into account the fact that all electrons in the conduction band do not have the lowest energy for that band and all the holes in the valence band do not have the highest valence band energy. As the temperature increases, electrons occupy states which are deeper into the conduction band and holes occupy states which are deeper into the valence band. More states come into play at higher temperatures than at lower temperatures. In this introduction, we neglect this modification to the two level model.

Problem 4. For silicon, the gap between the valence and conduction bands is about 1.78×10^{-19} J wide. Take $N_c = 2.88 \times 10^{25}$ m^{-3} and $N_v = 1.05 \times 10^{25}$ m^{-3}.

a. Modify the binary search program of Fig. 2-3 so that it can be used to solve Eqn. 28-6 for E_F. Write instructions so that, once E_F is calculated, the machine then calculates the density of electrons and the density of holes, using Eqns. 28-4 and 28-5, respectively.

b. For each of the following values of the temperature, calculate the Fermi energy, the density of electrons in the conduction band, and the density of holes in the valence band. Measure energies from the top of the valence band. That is, take $E_v = 0$ and $E_c = 1.78 \times 10^{-19}$ J.

    i. $T = 75^\circ$ K.

    ii. $T = 150^\circ$ K.

    iii. $T = 200^\circ$ K.

    iv. $T = 500^\circ$ K.

    v. $T = 1000^\circ$ K.

c. Plot E_F as a function of T. Extrapolate the curve and read the value of E_F for $T = 0^\circ$ K. Compare the value with the energy at the midpoint of the gap.

d. Make a graph of $\ln(n)$ as a function of $1/kT$. It is usual to present data in this form rather than in the form of n as a function of T because of the nearly linear nature of the curve which results. Measure the slope $\Delta \ln(n)/\Delta(1/kT)$ and compare the result with half the gap width. Save the graph for later use.

The extrapolation of part c shows that, at $T = 0^\circ$ K, the Fermi energy is at the center of the gap. At that temperature, all the states in the valence band are filled and all the states in the conduction band are empty.

If the Fermi energy were to remain at the midpoint of the gap as the temperature is raised, then the fraction of unoccupied states in the valence band would be the same as the fraction of occupied states in the conduction band. See the result of problem 1.

For silicon, however, the number of states in the valence band is less than the number of states in the conduction band so, if the density of electrons is to be the same as the density of holes, the fraction of unoccupied states in the valence band must be greater than the fraction of occupied states in the conduction band. The Fermi energy must move to below the midpoint of the gap. See the results of problem 3.

The density of electrons in the conduction band and the density of holes in the valence band play important roles in determining the electrical conductivity of the material. If an electric field $\vec{E}$ is turned on, the current density $\vec{j}$ is given by

$$\vec{j} = (env_e + epv_h)\vec{E}, \tag{28-8}$$

where e is the magnitude of the charge on the electron, v_e is the magnitude of the average velocity of the electrons in the conduction band, and v_h is the magnitude of the average velocity associated with the holes in the valence band.

For a given electric field, the current is strongly dependent on the temperature. The speeds v_e and v_h are temperature dependent but, over a wide range, it is chiefly the temperature dependence of the densities n and p which determines how the current changes with temperature.

As the temperature increases, more electrons are promoted from the valence band to the conduction band and the current increases. The quantity in the parentheses of Eqn. 28-8 is the conductivity of the material and, to find the gap width, it can be measured as a function of temperature, then its natural logarithm is plotted as a function of 1/kT. Since it follows the densities, the graph is linear and the slope has a magnitude which is half the gap width. This is done at high temperatures. At low temperatures, the variation of the electron and hole speeds with temperature must be taken into account.

The density of electrons in the conduction band and the density of holes in the valence band can be changed dramatically by the addition of certain kinds of impurities to the semiconducting material. The semiconductor is then said to be doped. In the following discussion, we assume that there is only one additional electron state associated with each impurity. This assumption is not correct since it neglects the spin angular momentum of the electron but, for the problems considered here, it simplifies the complexity of the problem considerably without drastically changing numerical results.

One type of impurity atom brings an additional state and an additional electron with it. The energy level of the state is in the gap between the valence and conduction bands, usually close to the conduction band. Such impurities are called donors since many of the electrons associated with them are thermally excited to the conduction band.

Another type of impurity atom brings an additional state, with energy in the gap, but does not bring an additional electron. Such impurities are called acceptors since electrons from the valence band are thermally excited to the impurity level, leaving behind holes in the valence band.

It is only the electrons in the conduction band and holes in the valence band, not electrons in impurity levels, which contribute to the electrical current when an electric field is turned on. Impurities increase the conductivity of the material by increasing the number of electrons in the conduction band or the number of holes in the valence band.

Suppose N_d donors per unit volume are added to the material. Each brings one additional state, with energy E_d. The density of electrons in donor levels is then given by

$$n_d = \frac{N_d}{e^{(E_d - E_F)/kT} + 1} \, . \tag{28-9}$$

The Fermi energy is again determined by the condition that, at a given temperature, the number of electrons, distributed over states in the two bands and the donor states, is equal to the total number of electrons in those states at $T=0^\circ$ K. At $T=0^\circ$ K the valence band is filled, the donor states are filled, and the conduction band is empty. The total density of electrons in the states of interest is $N_v + N_d$ and

$$\frac{N_c}{e^{(E_c - E_F)/kT} + 1} + \frac{N_v}{e^{(E_v - E_F)/kT} + 1} + \frac{N_d}{e^{(E_d - E_F)/kT} + 1} = N_v + N_d . \tag{28-10}$$

The first term on the left gives the density of electrons in the conduction band, the second term gives the density of electrons in the valence band, and the third term gives the density of electrons in the donor states.

It is worthwhile to carry out the subtractions $N_v - N_v / \left[e^{(E_v - E_F)/kT} + 1 \right]$ and

$N_d - N_d / \left[e^{(E_d - E_F)/kT} + 1 \right]$ algebraically and write

$$\frac{N_c}{e^{(E_c - E_F)/kT} + 1} - \frac{N_v}{e^{-(E_v - E_F)/kT} + 1} - \frac{N_d}{e^{-(E_d - E_F)/kT} + 1} = 0. \qquad (28\text{-}11)$$

This equation is to be solved for E_F, then Eqns. 28-4, 28-5, and 28-9 are evaluated for the density of electrons in the conduction band, the density holes in the valence band, and the density of electrons in the donor states, respectively.

Problem 5.   Phosphorous atoms in silicon act as donors, with an impurity level about 1.71×10^{-19} J above the top of the valence band. The gap between the conduction and valence bands is about 1.78×10^{-19} J wide. Take $N_c = 2.88 \times 10^{25}$ m^{-3} and $N_v = 1.05 \times 10^{25}$ m^{-3}.

a.   Modify the binary search program so that it can be used to solve Eqn. 28-11 for E_F.
     Write instructions so that, once E_F is found, the machine evaluates n, p, and n_d.

b.   Suppose 1×10^{12} phosphorous atoms per m^3 are added to silicon. They each bring one
     additional electron and one additional state in the gap. For each of the following
     temperatures, calculate the Fermi energy, the density of electrons in the conduction
     band, the density of holes in the valence band, and the density of electrons in the
     impurity levels.
     i.   T=75° K.
     ii.   T=150° K.
     iii. T=250° K.
     iv.   T=300° K.

c.   Suppose 1×10^{18} phosphorous atoms per m^3 are added to pure silicon. For each of the
     following temperatures, repeat the calculations of part c.
     i.   T=75° K.
     ii.   T=300° K.
     iii. T=500° K.
     iv.   T=1000° K.

d.   For each impurity density considered, plot $\ln(n)$ as a function of $1/kT$. Make both
     these plots on the graph you made in response to problem 4.

       For both of the donor densities considered, the qualitative features are the same.
At the lower temperatures, the density of electrons in the conduction band is nearly the

same as the density of donors and the donor levels are nearly depleted. The impurities have donated their electrons to the conduction band.

Notice also that the density of holes in the valence band is quite small. In fact, it is smaller than the density for the pure material at the same temperature. The Fermi energy has shifted to a value well above the midpoint of the gap.

Although you have not investigated this aspect, it is true that, as long as the number 1 can be neglected in Eqns. 28-4 and 28-5, the product np does not depend on the value of the Fermi energy. At the same temperature, the product is the same for the material with impurities as it is for the pure material. When donors are added, the density of electrons in the conduction band increases and the density of holes in the valence band decreases, but the product remains the same.

If the temperature is lowered to values below those considered in the problem, the Fermi energy moves to a value above the donor level. That this must be so is easily understood when it is remembered that, at $T=0^{\circ}$ K, the donor levels are filled and the Fermi energy is above the highest filled state. As the temperature is raised from 0° K, electrons from the donor levels are thermally excited to the conduction band. The energy needed to do this is much less than that needed to excite electrons from the valence band.

At a fairly low temperature, nearly all the electrons from the donor levels have been promoted. Since the temperature is still too low to excite many electrons from the valence band, the density of electrons in the conduction band is constant with temperature and its value is nearly the same as the density of donors.

At higher temperatures, the number of electrons from the valence band becomes comparable to the number excited from donor levels and n starts to increase with T. Notice that, in this temperature region, p also increases with T.

At still higher temperatures, the number of electrons excited from the valence band is much larger than the number excited from the donor levels and n is nearly the same as for the pure material. Now n and p are almost the same. As the temperature increases, the Fermi energy drops back toward the value it would have if the material were pure.

We now turn to a study of acceptor impurities in an otherwise pure semiconductor. If N_a acceptor impurities per unit volume are added to pure semiconducting material, the density of electrons in acceptor states is given by

$$n_a = \frac{N_a}{e^{(E_a - E_F)/kT} + 1} , \qquad (28\text{-}12)$$

where E_a is the acceptor energy level. It is assumed that each acceptor atom brings one additional state, with energy in the gap between the valence and the conduction bands.

The Fermi energy is now the solution to

$$\frac{N_c}{e^{(E_c - E_F)/kT} + 1} - \frac{N_v}{e^{-(E_v - E_F)/kT} + 1} + \frac{N_a}{e^{(E_a - E_F)/kT} + 1} = 0. \qquad (28\text{-}13)$$

The acceptor atoms bring no additional electrons so the total number of electrons, distributed among the three sets of states, is N_v.

Problem 6.   Galium atoms in silicon act as acceptors and the impurity level is about 1.04×10^{-20} J above the top of the valence band. The gap between the valence and conduction bands is about 1.78×10^{-19} J. Take $N_c = 2.88 \times 10^{25}$ m^{-3} and $N_v = 1.05 \times 10^{25}$ m^{-3}.

a.   Modify the binary search program so that it can be used to solve Eqn. 28-13 for the Fermi energy. Write instructions to calculate the density of electrons in the conduction band, the density of electrons in the acceptor states, and the density of holes in the valence band.

b.   Suppose 1×10^{16} galium atoms per m^3 are added to pure silicon. For each of the following temperatures, calculate the Fermi energy, the density of electrons in the conduction band, the density of electrons in the acceptor states, and the density of holes in the valence band.

   i. $T = 75^\circ$ K.

   ii. $T = 150^\circ$ K.

   iii. $T = 300^\circ$ K.

   iv. $T = 500^\circ$ K.

c.   On the graph you made in response to problem 4, plot $\ln(n)$ vs. $1/kT$.

d.   Qualitatively explain the behavior of the curve in terms of the excitation of electrons and the shift in the Fermi energy.

The basic building block of many solid state electronic devices is the p-n junction. A p-n junction is constructed so that, on one side of the junction, the semiconducting material is doped with acceptors while, on the other side, the material is doped with donors. The acceptor side is labelled p and the donor side is labelled n.

Suppose a p-n junction is formed by joining some p type silicon to some n type silicon. When the junction is first formed, some electrons from the n side diffuse to the p side and fall into vacant states in the valence band. Likewise, some holes from the p side diffuse to the n side where electrons drop from the conduction band to fill some of the vacant states. On both sides of the junction there is a net charge, which resides on the impurities. The n side is positively charged because donors there have lost electrons to the p side. The p side is negatively charged because acceptors there have lost holes to the n side.

As the electrons and holes diffuse across the boundary, the charged impurities left behind produce an electric field in the neighborhood of the boundary. Since the n side is positively charged, this side is at a higher electric potential than the p side and electron energies on the n side are lowered relative to those on the p side.

If E_c and E_v are the energies of the conduction and valence bands, respectively, on the p side, then E_c-eV and E_v-eV are the energies of the conduction and valence bands, respectively, on the n side. Here V is the potential difference and these expressions result when the potential energy $-eV$ due to the electric field is added to the value of each electron energy level. Only potential differences are physically significant and we have arbitrarily chosen to add the potential energy to the levels on the n side rather than subtract it from the levels on the p side.

The electric field is directed from the n side toward the p side and is in the proper direction to inhibit the flow of both electrons and holes. As the field builds up, an equilibrium situation develops and, when that condition is reached, the flow stops. The electric field and the potential difference exist when equilibrium is reached, even though there is no net flow of charge then.

At equilibrium, the whole sample, including both sides of the junction, is characterized by a single Fermi energy. All parts of the material are in thermodynamic equilibrium at the same temperature. Before the junction is made, the Fermi energy for the p type material is in the lower half of the gap while the Fermi energy for the n type material is in the upper half of the gap. After the junction is formed, the energy

578

levels and the Fermi level on the n side are lowered, relative to those on the p side, by the electric field. They are lowered just enough for the Fermi energy to be the same on the two sides. Then no current flows. This process is analyzed quantitatively in the next problem.

Problem 7.   A p-n junction is formed by two pieces of silicon. On the p side, there are 1×10^{16} galium atoms per m^3 and, on the n side, there are 1×10^{18} phosphorous atoms per m^3. Use the results of problems 5 and 6.

a.   Draw energy level diagrams, side by side and to scale, for the n side and for the p side, before the junction is made. Put the two conduction bands at the same level and the two valence bands at the same level. The gap, of course, is the same for the two materials. Also indicate the positions of the Fermi energies on the diagram.

b.   By how much should the levels on the n side be lowered so that the Fermi energies are at the same level on the two sides? How large an electric potential drop from the n side to the p side accomplishes this lowering?

c.   Calculate the ratio of the electron density on the n side to the electron density on the p side. Compare the ratio with e^{eV}, where V is the electric potential drop found in part b. Also calculate the ratio of the hole density on the p side to the hole density on the n side and compare it with the same quantity.

        The potential drop is called the contact potential. A contact potential exists at the junction whenever two dissimilar materials are joined. As in the situation presented in problem 7, it results from the flow of charge from one side to the other.

        A junction diode consists of a single p-n junction. Current flows easily in one direction, from the p to the n side, when a power supply is connected with its positive terminal attached to the p side. When the power supply leads are reversed, very little current flows.

        If an external power supply is attached, with the positive terminal connected to the p side and the negative terminal connected to the n side, the energy levels on the n side are raised relative to those on the p side and electrons flow from the n side to the p side. Holes also flow from the p side to the n side. Since the populations of the states are nearly exponential functions of the energy, a small external voltage can produce a large flow of charge across the junction.

If the leads on the power supply are reversed, so the negative terminal is connected to the p side, the levels on the n side are lowered relative to those on the p side. Electrons do flow from the p side to the n side, but there are relatively few electrons on the p side and the current is small.

28.2   Radioactivity

An atomic nucleus consists of a collection of tightly bound neutrons and protons. Nuclei are commonly designated by the atomic number Z, which is equal to the number of protons, and the atomic weight A, which is equal to the total number of nucleons (protons and neutrons). The usual symbol for a nucleus has the form

$$_{Z}S^{A}$$

where S is the chemical symbol for the nucleus. The atomic weight is written as a super-script after the chemical symbol and the atomic number is written as a subscript before the chemical symbol. Since the chemical symbol is sufficient to specify the atomic number, Z is often omitted.

Two nuclei with the same atomic number but different atomic weights are called isotopes. They have the same number of protons but differ in the number of neutrons. The atoms are chemically similar but the nuclei may be quite different in their properties. $_{8}O^{16}$ and $_{8}O^{17}$, for example, are two isotopes of oxygen.

Many nuclei are unstable and undergo transitions, usually including the emission of particles, to achieve stability. These nuclei are said to be radioactive and the transitions are called radioactive decays. There are six distinct decay processes which are common.

α emission.   A helium nucleus, called an α particle, is emitted and the daughter nucleus has an atomic weight which is 4 less than that of the parent nucleus and an atomic number which is 2 less than that of the parent nucleus. An example is

$$_{92}U^{238} \rightarrow \ _{90}Th^{234} + \ _{2}He^{4} .$$

In this example, a uranium nucleus, with 92 protons and 146 neutrons (238 nucleons all together) decays to a thorium nucleus, with 90 protons and 144 neutrons.

In α decay, the number of protons and the number of neutrons are separately conserved. Two of each of these particles, formerly in the parent nucleus, are emitted but retain their identity in the helium nucleus.

The sum of the rest energies of the daughter nucleus and the helium nucleus is less than the rest energy of the parent nucleus. The remaining energy appears as kinetic energy of the decay products, with a greater fraction going to the helium nucleus.

β emission. Either an electron, called a β^- particle, or a positron, called a β^+ particle, is emitted. The β particle is always accompanied by another particle, either a neutrino or an antineutrino.

In β^- decay, a neutron in the parent nucleus is converted to a proton with the emission of an electron and an antineutrino. The atomic weight of the daughter nucleus is the same as that of the parent but its atomic number is greater by 1. An example is

$$_6C^{14} \rightarrow {}_7N^{14} + e^- + \bar{\nu}$$

where $\bar{\nu}$ represents the antineutrino. A neutron in the carbon nucleus changes to a proton and emits an electron and an antineutrino. With one more proton and one less neutron, the nucleus is now a nitrogen nucleus.

In a β^+ decay, a proton in the parent nucleus is converted to a neutron. A positron and a neutrino are emitted. The atomic weight of the nucleus remains the same but its atomic number is reduced by 1. An example is

$$_{11}Na^{22} \rightarrow {}_{10}Ne^{22} + e^+ + \nu$$

where ν represents the neutrino. A proton in the sodium nucleus emits a positron and a neutrino and changes to a neutron. The resulting nucleus is an neon nucleus.

Most of the energy lost by the nucleus appears as the rest and kinetic energies of the β particle and the kinetic energy of the neutrino. It is believed that neutrinos, like photons, have zero rest energy. Nearly all the kinetic energy is shared by the two light particles and there is no restriction on the fraction received by the β particle.

Electron capture.   In this process, an electron from outside the nucleus enters the nucleus and interacts with one of the protons. As a result, the proton becomes a neutron and a neutrino is emitted. An example is

$$_{20}Ca^{41} + e^- \rightarrow {}_{19}K^{41} + \nu.$$

The electron is usually from a low lying atomic state. Following the capture, electrons in higher energy states fall to vacated lower states until the lowest energy electron configuration is reached. X-ray radiation is emitted during this rearrangement process.

In both positron emission and electron capture, a proton changes into a neutron and the atomic number of the nucleus is reduced by 1. The two processes are competitive. Most isotopes which undergo one of these types of transition also undergo the other.

Fission.   Some heavy nuclei decay by breaking into two lighter nuclei, both in the midrange of atomic weights for naturally occuring nuclei. Such a process is called fission. An example is

$$_{92}U^{236} \rightarrow {}_{39}Y^{95} + {}_{53}I^{139} + 2n,$$

in which a uranium nucleus splits into yttrium and iodine nuclei. Two neutrons are also emitted.

Spontaneous fission is rare and most fission events are induced by bombarding a heavy target nucleus with a particle, usually a neutron. For example,

$$n + {}_{92}U^{235} \rightarrow {}_{39}Y^{95} + {}_{53}I^{139} + 2n.$$

The result of $_{92}U^{236}$ fission, or any other fission process, is not always the same two elements, but may be any one of a large number of combinations of possible fragments.

The sum of the rest masses after the fission event is less than the sum before. The difference appears as the kinetic energies of the fragments and the neutrons.

γ emission.   In this process, a nucleus drops from a higher energy state to a lower one
and the energy is carried away by a high energy photon, called a γ particle. Both the
atomic weight and the atomic number of the nucleus remain the same as they were before
the emission.   Only the internal energy and the state of motion of the nucleons change.

In many situations, one of the other decay processes leaves the daughter nucleus
in an excited state and the daughter decays, via γ emission, to its lowest energy state.

Internal conversion.   This process competes with γ emission. The internal energy,
instead of going to a photon, is given to one of the orbital electrons. The electron is
ejected and x-ray radiation is produced as the other electrons fall into vacant states
and eventually arrive at the lowest energy electron configuration.

The decay of a single nucleus is a random event. The time of its occurrence
cannot be predicted. In a large collection of identical radioactive nuclei, some will
decay in any time interval selected, whether shortly after preparation of the sample,
long after, or any interval in between.

If the sample contains a sufficiently large number of nuclei, then equal fractions
of undecayed nuclei decay in equal time intervals. The number which decay in the
infinitesimal interval dt at time t is proportional to dt and to the number of undecayed
nuclei remaining at that time. If N(t) represents the number of nuclei which have not
yet decayed by time t, then

$$dN = -\lambda N(t) \, dt \qquad\qquad (28\text{-}14)$$

where λ is a constant, called the decay constant. It is positive; the negative sign in
the equation indicates that, as time goes on, the number of undecayed nuclei remaining
decreases.

This expression can be integrated numerically. A technique similar to that used
to integrate Newton's second law is used. Write dN/dt=f, where f is a function of N and
t. Divide the t axis into intervals, each Δt wide. If N_i is the value of N at the end
of interval i, then

$$N_{i+1} = N_i + f \, \Delta t \qquad\qquad (28\text{-}15)$$

is the value of N at the end of interval i+1. This expression is used repeatedly to obtain N at the end of a series of intervals, starting with the value of N for some initial time.

A flow chart is shown in Fig. 28-2. The storage allocation is

X1: N,
X2: t,
X3: Δt,
X4: n,
X5: counter,

and   X6: f.

Here n is the number of intervals to be considered before results are displayed. The program is started by placing the initial number of nuclei in X1, the initial time in X2, the interval width in X3, and n in X4. Since $\ln(N)$ is called for in many problems, an instruction to calculate it has been incorporated into the program, at line 200. You may wish to add instructions to calculate the activity, given by λN and discussed later.

The function f must be supplied at the time the machine is programmed. To integrate Eqn. 28-14, the instruction at line 140 should read

$$(-\lambda)*X1 \rightarrow X6$$

where a numerical value must be supplied for the decay constant λ.

Problem 1.   The aluminium nucleus $_{13}Al^{26}$ decays to magnesium via β^+ decay. The decay constant is 0.099 s^{-1}. Suppose there are 1×10^{19} $_{13}Al^{26}$ nuclei in the sample at time t=0.
a.   Use the program of Fig. 28-2, with Δt=0.05 s, to find the number of undecayed $_{13}Al^{26}$ nuclei at the end of every 2 s interval from t=0 to t=16 s. You should obtain three figure accuracy. To obtain more accuracy, reduce Δt.
b.   Plot the natural logarithm of N as a function of time. Draw a straight line through the points and determine the slope $\Delta(\ln N)/\Delta t$. Compare the result with λ.
c.   Use the graph to determine the time for which half the $_{13}Al^{26}$ nuclei have decayed.
d.   Use the graph to determine the time for which three fourths of the original $_{13}Al^{26}$ nuclei have decayed.

584

Figure 28-2.   Flow chart for a program to integrate the radioactive decay law.

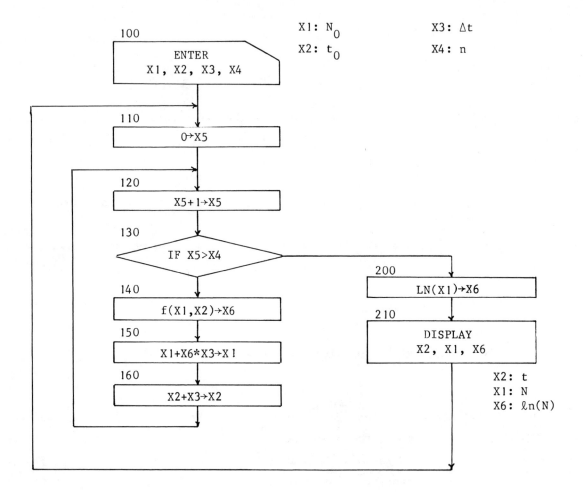

The natural logarithm of N is a linear function of t and the slope is $-\lambda$.

$$N(t) = N_0 e^{-\lambda t},\qquad(28\text{-}16)$$

where N_0 is the number of nuclei present at time t=0, is the exact solution to Eqn. 28-14. The decay constant can be found from experimental data by plotting $\ln N$ vs. t, then measuring the slope of the line.

$N(t)/N_0 = \frac{1}{2}$ when $t=\ln 2/\lambda$. This time is called the half life of the decay and it is denoted by T. It is the time during which half the original nuclei undergo decay. Half the remaining nuclei undergo decay in an additional time T so, at time t=2T, one fourth of the original nuclei remain undecayed. At time t=3T, one eighth of the original nuclei remain undecayed and so on.

The average lifetime of a nucleus is denoted by τ. It is calculated by summing the times of decay for the nuclei, then dividing by the number of nuclei originally present. The number of nuclei which decay in the time interval dt at time t is given by

$$|dN| = \lambda N(t)\, dt\qquad(28\text{-}17)$$

and these have all lived for time t without decaying. So

$$\tau = (1/N_0)\int_0^\infty t\lambda N(t)\, dt$$
$$= \lambda \int_0^\infty t\, e^{-\lambda t}\, dt.\qquad(28\text{-}18)$$

This integral can be evaluated analytically and the result is

$$\tau = 1/\lambda.$$

We shall evaluate it numerically in the next problem.

The average lifetime can be found graphically by plotting N as a function of time, then drawing the straight line tangent to the curve at t=0. Since the slope of N(t) is $-\lambda N_0$ at t=0, the equation for this line is

$$N = N_0(1 - \lambda t)\qquad(28\text{-}19)$$

and its intercept on the t axis is at t=1/λ, the average lifetime.

Problem 2.   Consider the β^+ decay of $_{13}Al^{26}$, as described in problem 1.   The decay constant is 0.099 s^{-1} and there are 1×10^{19} nuclei present at time t=0.

a.   Use the program of Fig. 8-1 to evaluate the integral in Eqn. 28-18. Replace the upper limit with 150 s and use 200 intervals. Compare the average lifetime with 1/λ.

b.   Plot N as a function of t, draw the line tangent to the curve at t=0, and find its intercept on the t axis. Use data generated in response to problem 1. Again compare the result with 1/λ.

In most cases, it is not possible to measure N(t) directly. It is possible, however, to measure a quantity which is proportional to the number of decays with occur in selected time intervals. This is the number of counts made, for example, by a Geiger counter. When the number of decays in a short time interval is divided by the time interval, the result is called the activity of the sample. For the simple decay discussed above, the activity R is given by $|dN/dt|$. Since $|dN/dt| = \lambda N(t)$, the activity is given by

$$R = \lambda N(t). \tag{28-20}$$

As we shall see, this expression is valid even in situations for which the activity is not the derivative of N.

For the decay discussed in problems 1 and 2, the natural logarithm of the activity is a linear function of time and its slope is $-\lambda$. The activity can be plotted as a function of time and the line tangent to the curve at t=0 drawn. This line intersects the t axis at t=τ, the mean lifetime. These conclusions all follow from the linear relationship between the activity and the number of nuclei.

Some samples contain more than one kind of radioactive nucleus. If that is the case, ℓnR is not a linear function of the time. The following problem gives an example.

Problem 3.   A sample contains 1×10^{21} phosphorous atoms. 4% are $_{15}P^{32}$, which decays according to

$$_{15}P^{32} \rightarrow _{16}S^{32} + e^+ + \nu$$

and has a half life of 14.3 days. 2% are $_{15}P^{33}$, which decays according to

$$_{15}P^{33} \rightarrow _{16}S^{33} + e^+ + \nu$$

and has a half life of 25 days. The rest are $_{15}P^{31}$ and are stable.

a. Find the decay constant, in d^{-1}, for each of the decays.

b. The activity of the sample is given by

$$R = 0.04N_0\lambda_1 e^{-\lambda_1 t} + 0.02N_0\lambda_2 e^{-\lambda_2 t}.$$

Use the program of Fig. 2-4 to draw a graph of the natural logarithm of the activity as a function of time, from t=0 to t=300 d. Plot points for every 20 d. This is what a plot of experimental data looks like for this sample. It is not linear.

c. At large values of t, however, the graph is linear. The short lived nuclei have mostly undergone decay and nearly all the activity is due to the nuclei with the longer half life and shorter decay constant. Measure the slope of this part of the graph and compare its value with the decay constant for the longer lived nuclei.

d. Draw the line which is tangent to the ℓn(R) curve at large values of t and extend it back to t=0. This is what ℓn(R) would look like if only the longer lived nuclei were present in the sample. For each of the first 10 values of t, starting with t=0, subtract the activity of the longer lived component from the total activity and plot the natural logarithm of the difference on the same graph. The result should be a straight line. It represents the natural logarithm of the activity as a function of time for the shorter lived component. Measure its slope and compare the value with the decay constant.

The above problem is a demonstration of a technique commonly used to analyze experimental data when two radioactive components are present in the same sample. It can sometimes be used when more than two components are present.

Because the half lives are nearly the same for the decays of problem 3, it was necessary to look at data for very large values of the time: 12 average lifetimes of the longer lived component. Had there been a greater difference in the half lives of the components, data could have been taken for a smaller number of average lifetimes.

The number of daughter nuclei can easily be found as a function of time. There are two cases of interest. In the first case, the daughter nucleus is stable while, in the second, it is radioactive.

If the daughter is stable, then the number dN_2 of daughter nuclei created in time dt is equal to the number of parent nuclei which decay in that time interval or

$$dN_2/dt = -dN_1/dt \qquad\qquad (28-21)$$

where N_1 is the number of undecayed parent nuclei present at time t. This equation can be integrated to give

$$N_2(t) = N_{10} - N_1(t) \qquad\qquad (28-22)$$

where N_{10} is the original number of parent nuclei. It is assumed that no daughter nuclei are present initially. Since $N_1(t)=N_{10}e^{-\lambda t}$, the last equation can be written

$$N_2(t) = N_{10}(1 - e^{-\lambda t}). \qquad\qquad (28-23)$$

If two types of parent nuclei are present, as in problem 3, each of them is treated separately. In the above equations, N_{10} is then the original number of parent nuclei of one type and $N_2(t)$ is the number of daughter nuclei produced by the decay of those nuclei.

Problem 4. Consider the decay of $_{13}\text{Al}^{26}$, described in problem 1. The decay constant is 0.099 s^{-1} and there are originally 1×10^{19} undecayed nuclei. The daughter nuclei are stable. Assume there are none present at t=0. Use the program of Fig. 2-4 to plot the number of daughter nuclei as a function of time. Plot points for every 2 s from t=0 to t=50 s.

The curve rises with t and approaches the value 1×10^{19} as a limit. In this limit, all of the parent nuclei have decayed and there is one daughter nucleus for each parent nucleus originally present.

At any time t, the ratio of the number of daughter nuclei to the number of parent nuclei is given by

$$N_2(t)/N_1(t) = (1 - e^{-\lambda t})/e^{-\lambda t} = e^{\lambda t} - 1. \qquad (28\text{-}24)$$

Eqn. 28-24 can be solved for the product λt:

$$\lambda t = \ln\left[\frac{N_2(t)}{N_1(t)} + 1\right]. \qquad (28\text{-}25)$$

This equation is used in two ways. First, it can be used to find λ from experimental data. N_2, N_1, and t are measured, then λ is calculated. Second, it can be used to find t if λ is known. If none of the nuclei have escaped the material, and if the material originally contained no daughter nuclei, then measurements of N_1 and N_2 can be used to find the time that has elapsed since the decay process started. This technique is used to date materials.

If the daughter nucleus is radioactive, the net increase in daughter nuclei during the time interval dt is the difference between the number produced by the decay of the parent and the number of daughters which decay, both in the time interval dt. That is,

$$dN_2 = -dN_1 - \lambda_2 N_2 \; dt, \qquad (28\text{-}26)$$

where λ_2 is the decay constant for the decay of the daughter. Since $dN_1 = -\lambda_1 N_1 \; dt$, Eqn. 28-26 can be written.

$$dN_2 = (\lambda_1 N_1 - \lambda_2 N_2) \; dt. \qquad (28\text{-}27)$$

Again the program of Fig. 28-2 can be used to perform the integration to obtain N_2 as a function of time. The function which appears at line 140 is $\lambda_1 N_1 - \lambda_2 N_2$. N_2 is in X1 and the program is started by placing the initial value there. N_1 must be known as a function of time.

When the daughter nuclei are simultaneously being produced and decaying, the activity is no longer taken to be $|dN_2/dt|$. The activity is defined so that it is a measure of the number of particles emitted by the radioactive nuclei and so it is taken to be the rate of decay only. That is, the activity of the daughter nuclei is given by $\lambda_2 N_2$.

In the next problem, the number of undecayed parent nuclei is held constant. As the nuclei decay, they are replaced by other undecayed parent nuclei, perhaps as the result of another radioactive decay process.

Problem 5.   A certain barium isotope has a half life of 300 hours and decays to lanthanum which, in turn, also decays. The half life of the lanthanum isotope is 40 hours.

a.   Calculate the decay constant for the barium nuclei and for the lanthanum nuclei.

b.   Use the program of Fig. 28-2 to calculate the number of undecayed lanthanum nuclei as a function of time. Use a time interval $\Delta t=1$ h and find N_2 at the end of every 10 h time interval from t=0 to t=120 h. Assume the number of barium nuclei remains constant at 1×10^{19} and that the number of lanthanum nuclei is initially zero.

c.   Plot the activity $\lambda_2 N_2$ of the lanthanum as a function of time.

d.   The activity is clearly approaching a limit. Put $\Delta t=5$h and continue the integration to 570 h. Compare the last value of the lanthanum activity with the barium activity.

The curve showing the lanthanum activity rises rapidly at first. Lanthanum nuclei are produced by the decay of barium and, since N_2 is small, only a few of them decay. As the number of lanthanum nuclei increases, more decay and an equilibrium situation is approached. At equilibrium, the number of lanthanum nuclei produced is equal to the number which decay in any time interval and so the number which exist at any instant remains constant. This condition is known as secular equilibrium.

According to Eqn. 28-27, $dN_2=0$ and equilibrium is reached when $\lambda_1 N_1=\lambda_2 N_2$. The barium activity is then the same as the lanthanum activity. In part d of problem 5, you checked this equilibrium condition.

A close approximation to secular equilibrium is reached if the number of parent nuclei is not artificially held constant but the lifetime of the parent is very long compared to that of the daughter. For example, the uranium isotope $_{92}U^{238}$, with a half life of 4.5×10^9 years, decays to a thorium isotope with a half life of 24.1 days. Over periods of time on the order of years, the number of uranium nuclei is essentially constant and the curve which shows the activity of the thorium nuclei as a function of time looks like the one you plotted for lanthanum in the last problem. Eventually, of course, the number of uranium nuclei decreases.

We now consider a situation in which the number of parent nuclei is not held constant but is allowed to decrease by radioactive decay. The number of parent nuclei is given by $N_1(t) = N_{10} e^{-\lambda_1 t}$ and this expression must be used in Eqn. 28-27.

As an example, consider the case for which the half life of the parent is longer than the half life of the daughter. At small values of the time, $\lambda_1 N_1$ is larger than $\lambda_2 N_2$, so N_2 increases. The increase continues until $\lambda_1 N_1 = \lambda_2 N_2$, then $dN_2 = 0$ and N_2 is at its maximum value. Thereafter, $\lambda_2 N_2$ is larger than $\lambda_1 N_1$ and N_2 decreases with time.

<u>Problem 6.</u>   Consider the decay chain of problem 5. A barium isotope, with a half life of 300 h, decays to lanthanum, with a half life of 40 h. The decay constants were calculated in response to part a of problem 5. At t=0, there are 1×10^{19} radioactive barium nuclei and no lanthanum nuclei present.

a.   Use the program of Fig. 28-2 to calculate the number of lanthanum nuclei as a function of time. Find the number at the end of every 20 h interval from t=0 to t=400 h. For the first 100 h, use $\Delta t = 1$, then, to save time, use $\Delta t = 5h$. In preparation for part b, also have the machine calculate the activity of the barium and the activity of the lanthanum for the values of the time given above. These are given by $\lambda_1 N_1$ and $\lambda_2 N_2$ respectively.

b.   On the same graph, plot the activity of the barium and the activity of the lanthanum, as functions of time.

c.   Use the graph to estimate the time for which the lanthanum has maximum activity. Verify that the activities of the parent and daughter are the same for that value of the time.

d.   For t=300 h and t=400 h, calculate the ratio of the lanthanum activity to the barium activity.

e.   Estimate the slope of $\ln(N_2)$ vs. t between t=300 h and t=400 h. Compare its value to λ_1 and to λ_2.

The behavior of the lanthanum activity for small values of t is much the same as it is for the situation of problem 5. As more lanthanum nuclei are produced than decay, the activity increases.

Now, however, the number of lanthanum nuclei produced per unit time decreases as the barium decays and the lanthanum activity reaches a peak. The peak occurs at the time when the two activities are the same. According to Eqn. 28-27, from then on, more lanthanum nuclei decay than are produced per unit time.

Eventually the two decay rates fall together. The number of lanthanum nuclei and the activity of lanthanum both decrease with time proportionally to $e^{-\lambda_1 t}$. The barium is now in control and the ratio of the activities is nearly constant. Lanthanum nuclei must be produced before they can decay. In fact, the activity ratio tends toward a limiting value given, in general, by $\lambda_2/(\lambda_2-\lambda_1)$. For the situation of problem 6, the limiting value is 1.154.

The results of the last problem should substantiate these conclusions. Remember, however, that there is some round off error and 400 h is not a long enough time to reach agreement with limiting values. If you wish to obtain better agreement, use a smaller Δt and allow the integration to run to a larger value of the time.

When the ratio of the activities is essentially constant and the two activities decrease together, the decays are said to be in transient equilibrium.

This situation is to be contrasted with the situation in which the lifetime of the daughter is longer than that of the parent. Then the parent nuclei decay rapidly to produce a collection of daughters. These then decay at a rate determined by their own decay constant, rather than by the decay constant of the parent.

Problem 7.   The ruthenium isotope $_{44}Ru^{105}$ has a half life of 4.5 h and decays to the rhodium isotope $_{45}Rh^{105}$. This, in turn, has a half life of 35 h and decays to palladium. Suppose there are 1×10^{19} $_{44}Ru^{105}$ nuclei and no $_{45}Rh^{105}$ nuclei present initially.

a.   Calculate the decay constants for the two decays. Use the program of Fig. 28-2 to calculate the number of rhodium nuclei present at the end of each 5 h interval from t=0 to t=100 h. Use $\Delta t=0.1$ h until t=20 h, then use $\Delta t=0.5$ h. Also have the machine calculate the activity of the ruthenium and the activity of the rhodium for each of times for which results are displayed.

b.   On the same graph, plot the activity of ruthenium and the activity of rhodium, as functions of time.

c.   For t=75 h and for t=100 h, calculate the ratio of the rhodium activity to the ruthenium activity.

No equilibrium condition, either secular or transient, is reached for this situation. The number of rhodium nuclei increases rapidly at first, as the ruthenium decays. Then the number of rhodium nuclei decreases with a half life of 35 h.

Since the activity of ruthenium decreases with time faster than the rhodium activity, the activity ratio grows with time. After a few half lives of the parent, nearly all the activity is due to the daughter nuclei.

Many nuclei have more than one mode of decay. For example, the copper isotope $_{29}Cu^{64}$ can decay, via β^- emission, to the zinc isotope $_{30}Zn^{64}$ or, via β^+ emission, to the nickel isotope $_{28}Ni^{64}$. About 38% of the $_{29}Cu^{64}$ nuclei present undergo the first type decay while the other 62% undergo the second type.

We consider a nucleus for which there are two modes of decay. A decay constant can be associated with each mode. If $N(t)$ represents the number of undecayed parent nuclei present at time t, $N_1(t)$ the number of daughter nuclei of one kind, and $N_2(t)$ the number of daughter nuclei of the second kind, then

$$dN_1 = \lambda_1 N \, dt \qquad (28\text{-}28)$$

and
$$dN_2 = \lambda_2 N \, dt. \qquad (28\text{-}29)$$

Here λ_1 is the decay constant for the first decay mode, λ_2 is the decay constant for the second decay mode, and we have assumed both daughters are stable.

The total change in N is the sum of the changes due to each mode of decay, considered separately, and so

$$dN = -(\lambda_1 + \lambda_2)N \, dt. \qquad (28\text{-}30)$$

This equation has the solution

$$N(t) = N_0 \, e^{-\lambda t} \qquad (28\text{-}31)$$

where $\lambda = \lambda_1 + \lambda_2$.

In general, the partial decay constants λ_1 and λ_2 are not known by direct experiment but they can be calculated if the fractional amounts of the various decay products are known. Suppose f_1 is the fraction of parent nuclei which decay via the first mode. In a time interval of duration dt, λN dt parent nuclei decay and $\lambda_1 N$ dt of them decay via the first mode. So.

$$f_1 = \lambda_1/\lambda, \qquad (28\text{-}32)$$

or

$$\lambda_1 = f_1\lambda, \qquad (28\text{-}33)$$

Similarly,

$$\lambda_2 = f_2\lambda, \qquad (28\text{-}34)$$

where f_2 is the fraction of parent nuclei which decay via the second mode.

Problem 8.   Consider the two decays

$$_{29}Cu^{64} \rightarrow {}_{30}Zn^{64} + e^- + \bar{\nu}$$

and

$$_{29}Cu^{64} \rightarrow {}_{28}NI^{64} + e^+ + \bar{\nu}.$$

The lifetime of $_{29}Cu^{64}$ is 12.8 hours and both decay products are stable. The first decay occurs for 38% of the parent nuclei and the second decay occurs for 62% of them. Suppose that, at t=0, there are 1×10^{19} $_{29}Cu^{64}$ nuclei present and none of the decay products have been created yet.

a.   Calculate the partial decay constants λ_1 and λ_2.

b.   Use the program of Fig. 28-2 to calculate the number of $_{30}Zn^{64}$ nuclei as a function of time. Use $\Delta t=0.2$ h and obtain values for the end of every 2 h interval from t=0 to t=30 h. The function used at line 140 is $\lambda_1 N_0 e^{-\lambda t}$.

c.   Use the program of Fig. 28-2 to calculate the number of $_{28}Ni^{64}$ nuclei as a function of time. Obtain values for the same times as in part b. The function used at line 140 is $\lambda_2 N_0 e^{-\lambda t}$.

d.   On the same paper, draw graphs of N, N_1, and N_2 as functions of time.

e.   For t=4 h, 6 h, 12 h, and 30 h, calculate the ratio of the number of $_{28}Ni^{64}$ nuclei to the number of $_{30}Zn^{64}$ nuclei.

f.   Both $N_1(t)$ and $N_2(t)$ are proportional to $1-e^{-\lambda t}$ where λ is the total decay constant. To show this, calculate $N_1/(1-e^{-\lambda t})$ and $N_2/(1-e^{-\lambda t})$ for t=6 h, 12 h, and 30 h. For each of the decay products, the result should be a constant but it is a different constant for the two types of daughters.

   As the decay progresses, the ratio of the numbers of product nuclei remains constant. It is always 0.62/0.38=1.63 for the decays considered. In each time interval, 62% of the decays go to $_{28}Ni^{64}$ and 38% go to $_{30}Zn^{64}$, so the ratio is maintained.

   For either decay, the number of daughter nuclei is proportional to $1-e^{-\lambda t}$ where λ is the total decay constant. This is to be contrasted with the situation described in problem 3. For that case, two different sets of parent nuclei are present and they decay independently of each other. One set of daughters increases in proportion to $1-e^{-\lambda_1 t}$ and the other set increases in proportion to $1-e^{-\lambda_2 t}$, each according to a different decay constant.

   For a single type of parent nucleus with two decay modes, a decay via one mode reduces the number of parent nuclei present and hence reduces the number of decays via the second mode in later time intervals.

Appendix A
TABLE OF RANDOM NUMBERS

The following numbers were selected randomly from a uniform distribution of integers from 0 to 1000. When a series of random numbers is required to work a problem, pick the first one from any place in the table. Then, to obtain successive numbers, continue down the column, using the numbers in the order in which they appear in the table. When the end of the column is reached, go to the top of the next column. If the last number in the table is used, continue by using the first.

543	653	325	943	326	335	645	538	354	67	697	355
22	233	494	1	312	147	512	720	495	174	110	480
382	841	251	873	666	335	276	172	531	486	385	30
654	89	122	184	567	363	751	461	287	733	412	938
784	76	209	254	10	395	664	780	61	53	77	939
837	574	811	26	622	121	85	111	262	396	446	75
390	842	550	746	337	602	581	262	988	646	962	582
908	236	428	744	146	643	654	636	967	152	542	331
976	673	554	855	136	400	722	149	657	45	211	549
477	971	781	577	111	260	602	235	516	81	724	283
124	429	728	182	22	27	899	444	269	656	424	349
488	634	605	446	96	916	184	146	864	601	408	258
763	338	457	363	284	984	594	749	890	705	913	51
236	729	315	850	559	888	166	510	601	709	667	222
617	658	870	339	604	717	716	553	110	644	438	761
266	729	731	192	6	329	16	457	689	768	220	676
255	183	990	765	112	646	764	971	850	627	666	492
262	627	347	878	814	108	181	860	659	688	69	430
161	634	925	380	486	419	557	130	708	919	104	151
212	802	93	740	953	201	292	931	356	684	42	929
39	543	297	458	344	761	68	893	371	109	22	294
674	821	101	891	88	111	823	659	940	50	152	489
661	183	725	111	881	348	216	360	927	857	613	454
739	428	770	154	895	443	110	517	42	67	193	382
201	837	66	221	59	901	338	364	614	352	990	785
835	157	258	119	488	368	417	535	66	709	858	783
731	642	345	194	702	787	829	926	796	945	512	355

```
143  621  734  988  167  544    4  903  931   19   73  638
196  852  831  286   54  607  723  765  747  695  696  968
608  447  427  128  290  184  940  619  964  812  780  589
865  110  601  171  652  187  767  332  991  165  896  820
299  808  287  197  386  412  866  156  367  561  936  924
507  925  535  508  982  716  613  541  623  209  145  650
344  131  228  797  865  127  488  675   78  714  298  508
136  666  292  789  592  134  302  141  201  516  474  654
463  468   56  534  354  204  481  666  778  729    5   80
300  581  256  367  462  234  184  368  862  291  783  815
881  490  974  338  169  378  956  795  951  259  946  239
497  785   75   35  558  535  782  286   81  522  796  185
389  174  917  632  774  172  727  487  822  217  126  465
390  557  506   37  612  579   16  131  250   26  256  198
583  298   39   78  286  263  604  316  408  133  212  459
 20  862  472  524  570   48  170  276  151  364  547  927
544  560  975  969   20  979  310  738  556  455  479   41
993  651  332  653  979  113  780  984  675  486  987  692
900  165  597  811  205  222  186   10  808  243  256   81
653  851  237  498  565  809  614  552  959  240  330  710
707  644  739  669  394  253  963  985  313  647  116  693
869  956  895  728  961  404  585  484  170  784  810  145
669  173   85  573  887  518  608  932  824  568  383  271
800    9  263  550  209  857  487   31  903  661  721  443
484  136   36  551  554  221  350  833  157  717   92  669
448  425  101  989  320   41  674  538  794   33   48  616
667  639  627  730  386  719  903   96  590  355  981  128
406   42  262  364  277  457  752  368  229  566  668  915
564  990  178  133  579  203  458  874   92  965  258  695
155  509  370  276  424  314   73  325  289  186  971  521
 74  110  215   44  518  673   45  157  338  526  206  982
374  997  938  311   36  654  551   77  699  811  547  424
835  691  753  808  629  587  570  617  298  946   51  853
183  708  943  170  747  308  359  241  407   35  933   76
526  274  991  473   19  462  173  156  373  439  705  486
624  989  152  781   47  519  435  278  642  777  244  418
478  413  926  678  797   14  255  419   60   66  660  321
165  661  880  334  632  271  698  108  818  971   19  334
218  295  520  485   21  999  381  217  622  925   10  438
```

20	624	179	39	907	997	923	318	281	420	693	169
678	447	338	684	889	126	578	179	994	903	289	213
367	425	542	107	845	357	40	719	713	702	556	321
498	476	415	684	774	919	164	622	61	708	858	10
303	882	708	12	44	475	684	46	12	532	904	632
452	620	227	619	234	540	662	41	392	199	733	47
123	586	275	215	74	303	307	941	116	746	519	26
740	509	240	717	658	366	174	797	517	288	436	32
481	597	427	399	172	249	828	36	461	770	643	661
188	659	602	882	366	509	187	432	744	247	10	936
53	912	141	250	499	404	991	964	75	556	483	333
117	334	174	581	590	731	888	809	398	468	990	488
526	745	422	917	815	829	797	598	773	946	239	234
219	425	168	323	656	357	956	41	600	32	919	748
652	741	900	176	565	516	391	292	795	752	9	115
664	559	510	62	180	462	799	612	791	923	888	963
327	623	301	135	297	58	410	349	257	59	482	322
964	22	820	933	249	424	16	159	547	699	725	610
958	811	134	734	255	216	927	551	11	766	881	43
998	166	211	274	945	396	158	495	891	974	297	349
536	30	480	741	988	731	808	551	681	571	76	618
392	217	951	801	896	797	128	864	432	343	139	14
423	451	627	598	290	434	209	328	395	114	95	442
792	880	926	369	903	369	414	75	705	792	545	408
505	769	123	267	788	570	990	277	527	224	539	77
383	765	317	959	686	766	589	573	770	66	834	836
190	24	783	980	775	784	329	696	240	359	147	943
806	656	567	357	340	872	996	477	412	563	661	714
987	196	972	496	166	756	841	278	954	867	303	131
453	724	676	362	293	115	791	187	832	376	149	295
339	76	281	438	844	376	36	463	13	166	428	725
671	387	559	636	246	638	603	556	496	116	784	404
483	532	698	396	912	795	488	707	817	640	802	531

Appendix B
A BRIEF SUMMARY OF THE FLOW CHART LANGUAGE

Memory locations are labelled with the letter X followed by a number: X1, for example, denotes memory location 1. The numbers in the storage locations may be used in arithmetic operations. The symbols +, -, *, /, and ↑ denote addition, subtraction, multiplication, division, and exponentiation, respectively. Once the operation has been performed, the result is placed in a memory location. Typical statements are

$$X1+X2 \rightarrow X3$$
$$X1+X2 \rightarrow X1$$
and
$$X1 \uparrow 2 \rightarrow X2$$

In the first case, the result of the addition is placed in memory 3 while, in the second case, it is placed in memory 1, thereby erasing the previous content of that memory. In the last example, the content of X1 is squared and the result is placed in X2.

If an instruction contains a string of arithmetic operations, exponentiation is performed first, then multiplication and division, and finally addition and subtraction.

$$X1*X2 \uparrow 2 \rightarrow X3$$

instructs the machine to multiply the square of X2 by X1 and place the result in X3.

$$X1+X2*X3 \rightarrow X1$$

instructs the machine to add X1 to the product of X2 and X3.

Parentheses are sometimes used to indicate the order in which operations are to be performed. The quantity inside inner parentheses is evaluated before the quantity inside outer parentheses.

Functions are written with the argument enclosed in parentheses.

$$X1*COS(X2) \rightarrow X3$$

instructs the machine to multiply the cosine of X2 by X1 and store the result in X3.

The machine should be programmed to follow the arrows shown in the flow charts. These may imply unconditional transfer statements, which instruct the machine to return to a previous instruction. This instruction must be labelled, usually by means of a number or letter, and the transfer statement refers to the label.

Conditional transfer statements have the form

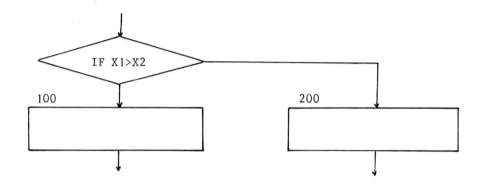

for example. If, at the time this instruction is encountered, the statement in the diamond shaped box is true, the machine next executes the instruction at line 200. If it is false, the machine next executes the instruction at line 100. For this example, if the number in X1 is greater than the number in X2, the machine goes to line 200, but if the number in X1 is less than or equal to the number in X2, the machine goes to line 100.

ENTER statements provide the means for placing numbers into memory locations from the keyboard or from other sources such as cards or tapes. A list of memory locations follows the word ENTER and the flow chart identifies the quantities to be placed in these locations.

When entry is from the keyboard, a STOP instruction must be included for each location on the list. The number is then keyed in and the machine restarted. The next instruction stores the number in the proper location, and the instruction after that stops the machine to receive the next number or gives the next instruction in the program.

When using hand held machines, it is often advantageous to omit ENTER statements from the program instructions and simply enter numbers manually into the correct storage locations before the program is run.

DISPLAY statements ask the machine to display the contents of the memory locations named in the list following the word DISPLAY. For hand held machines, this simply means recalling the number from memory, then stopping the program. Once the number has been copied, the machine is restarted manually. DISPLAY statements may be replaced by PRINT statements if the machine has printing capability. It is then not necessary to stop the machine.

More information about the flow chart language used in this book is given in Chapter 1 of PHYSICS PROBLEMS FOR PROGRAMMABLE CALCULATORS: MECHANICS AND ELECTROMAGNETISM. The appendices of that book contain specific instructions for translating the flow chart language to keystrokes for various widely used machines.

PROGRAMS FROM
PHYSICS PROBLEMS FOR PROGRAMMABLE CALCULATORS:
MECHANICS AND ELECTROMAGNETISM

Four programs from the previous volume are used in this book and, for each one, a flow chart and a brief description is given in this appendix. For more details, the reader is referred to the first volume. The figure numbers have been retained for easy reference.

C.1   Method of Uniform Intervals (Fig. 2-2)

This program is used to find the roots of a function $f(t)$. The range from t_0 to t_f is searched by dividing it into N intervals of equal length, then testing the sign of the function at the end of each interval. If the sign is different at the beginning and end of any interval, the end points of that interval are displayed, at line 410. After restarting, the machine resumes the testing of intervals.

If the sign is not different, the machine immediately goes on to the next interval. After all intervals in the range have been tested, the machine goes to line 300 where 1×10^{60} is displayed as a signal.

The limits of the search are entered at line 100, with t_0 in X1 and t_f in X2. The number of intervals N is placed in X3.

The function is evaluated at lines 120 and 230. Instructions must be supplied by the user and the value of the function must be placed in the proper memory location, as given on the flow chart.

The program fails to find a root if the function changes sign an even number of times in an interval or if the function passes through zero without changing sign. In the first case, the root can be found if N is made larger so that shorter intervals are used.

The larger the number of intervals used, the higher the accuracy with which the root is found. Increasing N, however, also increases the running time.

Figure 2-2.   Flow chart for the method of uniform intervals.

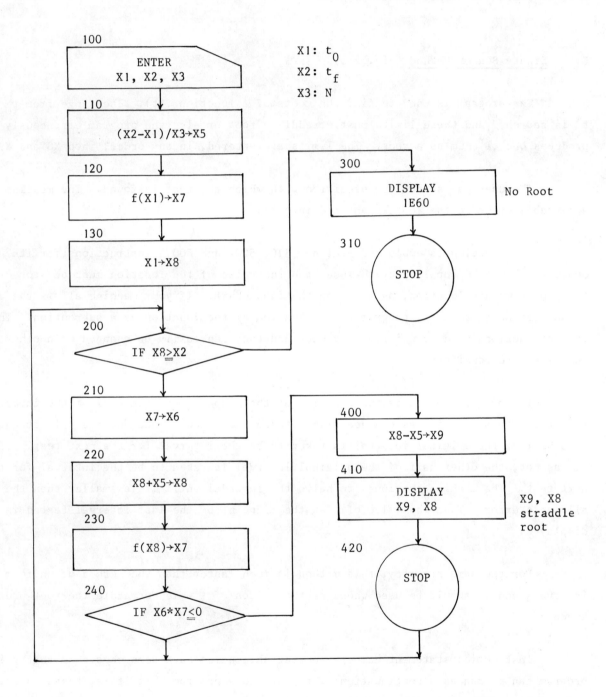

This program is of value when it is not known if the selected range contains a root. In those cases, it is often used as a preliminary step, prior to using the binary search method. It finds two values of t which straddle a root and these values are then used to start the binary search program.

C.2   Binary Search Method (Fig. 2-3)

The program is used to find the roots of a function $f(t)$. The range from t_0 to t_f is searched and these limits must straddle a root or else the program erroneously produces one of them as a root. The limits are entered, in any order, into X8 and X9.

The user can select the precision with which the root is found. The maximum acceptable error in the root is entered into X1.

The function is evaluated at lines 510, 520, and 600. Instructions for its evaluation must be supplied by the user and the value of the function must be stored in the proper memory location, as given on the flow chart. If your machine allows the use of subroutine statements, program the evaluation of the function as a subroutine. Then the instructions need be written only once and they can easily be changed when other functions are considered.

The program evaluates the function at the midpoint and the ends of the interval, then tests to see if it has changed sign between the midpoint and one end. If it has, this half of the original interval is taken to be the interval for the next test. If it has not, the other half of the original interval is taken to be the interval for the next test. The machine continues to halve the interval until it is smaller than the allowable error. Then the value of t at the midpoint of the last interval tested is displayed.

For the same accuracy, this method is much faster than the method of uniform intervals and it should be used whenever two values of t which straddle the root are known.

Instruction statement numbers and variable names have been chosen so that this program can be run as a continuation of the previous program. If it is, these instructions follow line 400 of that program and lines 410 and 420 are omitted. Only the maximum allowable error is entered at line 500.

605

Figure 2-3.   Flow chart for the binary search method.

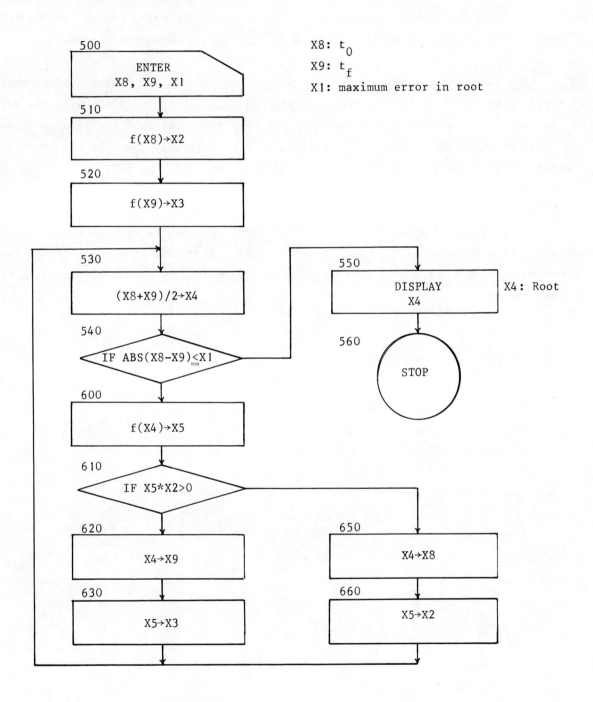

C.3 Plotting Program (Fig. 2-4)

This program evaluates a function f(x) at a sequence of values of the independent variable x. The initial value of x is entered into X1, the final value is entered into X2, and the increment is entered into X3. The function may also depend on two parameters and values for these are entered into X4 and X5 respectively. The number of parameters, of course, may be increased or decreased according to need.

The function is evaluated at line 120. Instructions for its evaluation must be supplied by the user. Once evaluated, its value is displayed, along with the value of the independent variable, at line 130. The STOP instruction is included to allow the user to copy the result. For some machines, a STOP instruction must be placed after the display of each number in the list. If results are printed, the STOP instruction may be omitted.

After displaying the value of the function, the machine returns to line 110 to consider the next value of x in the sequence. When all values of x in the given range have been considered, the machine goes to line 100, where new values of the parameters may be entered.

607

Figure 2-4. Flow chart for a program to evaluate a function for plotting.

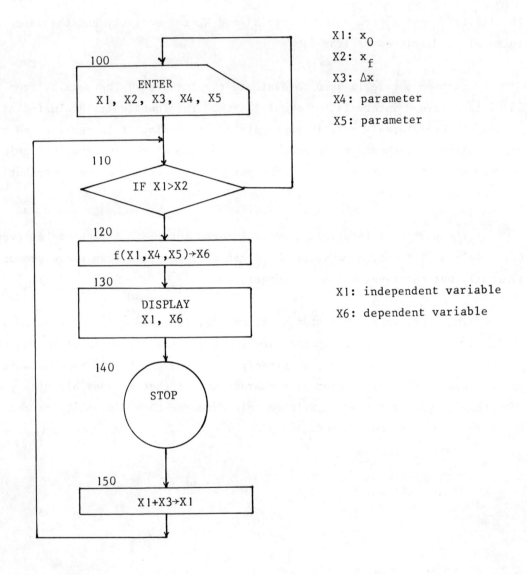

X1: x_0

X2: x_f

X3: Δx

X4: parameter

X5: parameter

X1: independent variable

X6: dependent variable

C.4   <u>Numerical</u> <u>Integration</u> <u>Program</u> (Fig. 8-1)

The program evaluates definite integrals of the form

$$\int_{x_0}^{x_f} f(x) \ dx.$$

The limits x_0 and x_f are entered into X1 and X2 respectively and the value of the integral is displayed at line 210.

Simpson's rule is used to evaluate the integral. The x axis, from x_0 to x_f, is divided into N intervals, of equal length. The function is evaluated at the end points of the intervals and, in each pair of intervals, it is represented by a second order polynomial, chosen to reproduce the function at the interval end points. The polynomial is then integrated over the pair of intervals and the contributions of all the interval pairs are summed.

The number of intervals used is entered into X3. It must be an even number. Large values of N produce answers which are more accurate than those produced when N is small, but the running time is longer.

The function is evaluated at lines 130, 170, and 180. Instructions for its evaluation must be supplied by the user and the value of the function must be placed in the proper memory location, as given on the flow chart. It is worthwhile programming the evaluation of the function as a subroutine, if that is possible with your machine. Then the instructions need be written only once and they can easily be changed when other functions are considered.

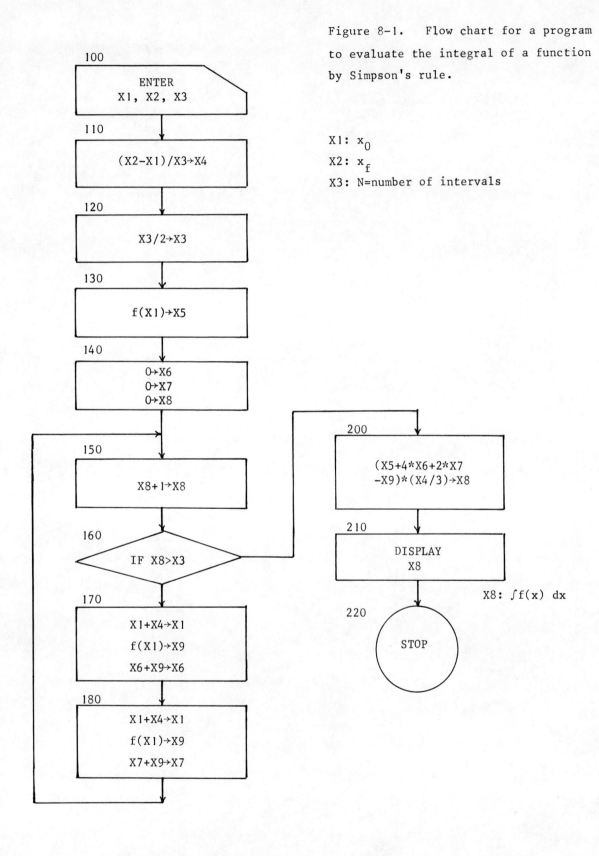

Figure 8-1. Flow chart for a program to evaluate the integral of a function by Simpson's rule.

X1: x_0
X2: x_f
X3: N=number of intervals

X8: $\int f(x)\ dx$